# The Importance of Quality

## Electrical Installation Series – Intermediate Course

M. Doughton

Edited by Chris Cox

THOMSON

Australia · Canada · Mexico · Singapore · Spain · United Kingdom · United States

**THOMSON**

The Importance of Quality

Copyright © CT Projects 2001

The Thomson logo is a registered trademark used herein under licence.

For more information, contact Thomson Learning, High Holborn House; 50-51 Bedford Row, London WC1R 4LR or visit us on the World Wide Web at:
http://www.thomsonlearning.co.uk

*British Library Cataloguing-in-Publication Data*
A catalogue record for this book is available from the British Library

ISBN 1-86152-714-4

**First published 2001 by Thomson Learning**
**Reprinted 2002, 2003 and 2005 by Thomson Learning**

Printed in Croatia by Zrinski d.d.

# About this book

"The Importance of Quality" is one of a series of books published by Thomson Learning related to Electrical Installation Work. The series may be used to form part of a recognised course or individual books can be used to update knowledge within particular subject areas. A complete list of titles in the series is given below.

# Electrical Installation Series

## Foundation Course

Starting Work
Procedures
Basic Science and Electronics

**Supplementary title:**
Practical Requirements and Exercises

## Intermediate Course

The Importance of Quality
Stage 1 Design
Intermediate Science and Theory

**Supplementary title:**
Practical Tasks

## Advanced Course

Advanced Science
Stage 2 Design
Electrical Machines
Lighting Systems
Supplying Installations

# Acknowledgements

The authors and publishers gratefully acknowledge the following illustration sources:

**ACAS** for Figure 1.8
**AEEU** for Figure 1.2
**AVO International Ltd (PowerSuite™)** for Figures 6.2, 6.3, 6.5, 6.7, 6.8, 6.12, 6.13, 6.14
**BSI** for Figure 6.1 Extracts from ISO9000, 2000 are reproduced with the permission of BSI under licence number 2000SK/0371. Complete British standards can be obtained by post from BSI Customer Services, 389 Chiswick High Road, London W4 4AL, United Kingdom. (Tel. UK + 020 8996 9001)
**CIP Ltd**, 60 New Coventry Road, Birmingham B26 3AY for Figure 8.3
**ECA** for Figure 1.3

**HMSO** for Figure 8.2 Crown copyright is reproduced with the permission of the Controller of Her Majesty's Stationery Office.
**JIB** for Figures 1.4, 1.6 and 1.7
**NICEIC** for Figure 1.5
**RS Components** for Figures 2.6, 2.7, 2.8, 2.9, 2.10, 2.11, 2.12 This product/publication includes images from CorelDRAW®9 which are protected by the copyright laws of the U.S., Canada and elsewhere. Used under license.

Every effort has been made to trace all copyright holders but if any have been inadvertently overlooked, the publishers will be pleased to make the necessary arrangements at the first opportunity.

# Study guide

This studybook has been written to enable you to study either in a classroom or in an open or distance learning situation. To ensure that you gain the maximum benefit from the material you will find prompts all the way through that are designed to keep you involved with the subject. The book has been divided into 26 parts each of which may be suitable as one lesson in the classroom situation. Certain parts of this book may be combined in one lesson period but this will depend upon the duration of the lesson. However if you are studying by yourself the following points may help you.

☞ Work out when, and for how long, you can study each week. Complete the table below and from this produce a programme so that you will know approximately when you should complete each chapter, the progress check and end test. Your tutor may be able to help you with this. It may be necessary to reassess this timetable from time to time according to your situation.

☞ Try not to take on too much studying at a time. Limit yourself to between 1 hour and 2 hours and finish if possible with a task or the self assessment questions (SAQ). When you resume your study go over this same piece of work before you start a new topic.

☞ You will find the answers to the questions at the back of the book but before you look at the answers check that you have read and understood the question and written the answer you intended.

☞ Multi-choice questions in a "Progress Check" at the end of Chapter 4 and an "End Test" covering all the material in this book are included so that you can assess your progress.

☞ Tasks are included where you are given the opportunity to ask colleagues at work or your tutor at college questions about practical aspects of the subject. These are all important and will aid your understanding of the subject.

☞ It will be helpful to have available for reference a current copy of "Health and Safety in Construction" HS(G) 150, BS 7671, IEE Guidance Note 3 and access to BS EN 60617 symbols for layout, circuit and wiring diagrams.

☞ Your safety is of paramount importance. You are expected to adhere at all times to current regulations, recommendations and guidelines for health and safety.

| Study times | a.m. from | to | p.m. from | to | Total |
|---|---|---|---|---|---|
| Monday | | | | | |
| Tuesday | | | | | |
| Wednesday | | | | | |
| Thursday | | | | | |
| Friday | | | | | |
| Saturday | | | | | |
| Sunday | | | | | |

| Programme | Date to be achieved by |
|---|---|
| Chapter 1 | |
| Chapter 2 | |
| Chapter 3 | |
| Chapter 4 | |
| Progress Check | |
| Chapter 5 | |
| Chapter 6 | |
| Chapter 7 | |
| Chapter 8 | |
| Chapter 9 | |
| End test | |

# Contents

# 1

# The Structure of the Industry and Planning the Installation

At this point at the beginning of each chapter you will be asked to complete an exercise to refresh your understanding of the topics covered in the previous chapter. These should be successfully completed before you progress to the next topic.

## On completion of this chapter you should be able to:

- identify the parties concerned with design and preparation of an installation
- state the importance and means of effectively establishing the clients' requirements
- state the importance of researching and collating relevant information
- state the importance of advising clients
- translate agreed requirements into an installation design
- state the importance of gathering the financial, legal, technical and resource information and the implications regarding its sufficiency and accuracy
- identify the organisations involved in labour relations in the electrical contracting industry
- identify the means of defining agreements, salaries and conditions within the industry
- state the means of dealing with disputes between employers and employees
- identify the contractual relationships which exist within the industry

## Part 1

## Labour relations

There are a number of bodies involved directly in the installation contracting industry. We shall consider the major ones, their place within the industry structure and how they relate to each other. Other organisations will be encountered and, whilst these will not be dealt with in detail, their functions will be similar to those given here as examples.

*Figure 1.1*

### AEEU

The Amalgamated Engineering and Electrical Union (AEEU), evolved from the original Electrical Trades Union which was formed in 1889.

The main function of the union is to safeguard the wages and working conditions of its members who are the employees of electrical contractors: electricians, apprentices and technicians.

*Figure 1.2      AEEU logo*

The union undertakes to represent its members in regard to their working conditions, salaries and holidays and to liaise with other trades unions. The union is not only concerned with protecting its members interests but also their physical safety and working conditions which have a direct effect on safety.

# E.C.A.

The Electrical Contractors Association was originally formed in 1901. It was given formal recognition by the board of trade in 1904. At that stage the Certificate of Incorporation did not allow it to negotiate in industrial matters. This became possible in 1919 by the forming of the National Federated Electrical Association which came under the trades union act of 1913. This enabled them to negotiate on conditions of employment and salaries in the electrical contracting industry.

*Figure 1.3      ECA logo*

In general terms the ECA is concerned with the employers' interests within the industry in much the same way as the union acts for the employee. The Association is governed by a council of elected members and it is set up with a region and branch structure throughout England, Ireland and Wales.

Terms and Conditions for the employer and the employee within the industry are negotiated with regard to their specific interests and as you can imagine these interests may not be the same.

# J.I.B.

From 1919 until 1968 the industrial relations within the industry were dealt with by the National Joint Industrial Council. This was a government-sponsored committee set up with the objective of securing peaceable industrial relations within the industry.

In 1968 the NJIC was taken over by the Joint Industry Board and was given the same government recognition. The JIB fulfils a large number of functions within the industry which we may summarise as a list:
• industrial relations
• employment
• grading
• health and safety
• training
• productivity
• welfare

*Figure 1.4      JIB logo*

The JIB rule book (1989) states "The principal objects of the JIB are.... for the purpose of stimulating and furthering the improvement and progress of the industry for the mutual advantage of the employers and employees engaged therein."

So the JIB is a combined body with representatives of both the AEEU and the ECA with an overall interest in the contracting industry. It must be said that not all electrical contractors belong to all or any of these bodies and that membership is not required to operate as a contractor.

# NICEIC

The National Inspection Council for Electrical Installation Contracting (NICEIC) is an independent body which was set up in 1956 by all sections of the electrical industry and it received charitable status in 1973.

*Figure 1.5      NICEIC logo*

The NICEIC registers Approved Contractors and regularly monitors their work. Its objective is to ensure the protection of the consumer of electricity throughout the British Isles, including the Channel Isles, the Isle of Man and Northern Ireland, against unsafe or defective electrical installations. A Roll of Approved Contractors is maintained and made available to the relevant areas of the industry and the public.

Inclusion on the NICEIC Roll is not compulsory and contractors may practise without ever becoming approved. The NICEIC is the industry's independent consumer protection body and it regularly assesses Approved Contractors to ensure they meet the requirements of the National standard, BS 7671 and other relevant Codes of Practice.

To summarise, the organisations that we have looked at so far are:

The AEEU is a body representing the employees' interests.

The ECA is a body representing the employers' interests.

The JIB is a body under which the above combine to determine the conditions within the industry as a whole.

The NICEIC is the only independent body representing the consumers' interest.

# Conditions of employment

Within the electrical contracting industry structure agreements, grading schemes and contracts of employment are produced. Each of these forms part of the conditions of employment which affect the majority of those involved in the industry. We shall take a look at these in turn and see how each affects the parties concerned.

## Agreements

These may be national agreements such as those for the installation of cables in buildings, ship work, onshore (as opposed to offshore) work and for specified engineering construction sites.

Contained within the agreement will be a number of clauses dealing with areas such as:
- the scope of work
- the type of employee
- the class of work
- demarcation
- method of working
- wages
- travelling and lodging allowances

There are also local agreements that may be formed to reflect local conditions. These will vary depending on the location and may cover any of the working conditions for that specific location.

It may be necessary to develop an agreement to cover a special site with its own particular requirements. A local agreement could be formed to determine the grades of labour that are to be used on site. For example in an area with a high percentage of unemployed, unskilled labour. In this case a local agreement may be drawn up to ensure that local labour is used for any unskilled work. This type of agreement is common when contracts are part of a redevelopment programme and, as part of the deal, employment for the local population is to be a priority. A further example could be a change in start or finish times, possibly extending the number of hours worked in a normal work period, to fit in with local shift working patterns.

# Grading schemes

Within any contracting industry it is likely that grading schemes will exist.

The Electrotechnical Certification Scheme was introduced, by the JIB, in June 1999 and is intended to replace the National Register of Electricians and identify the holders of JIB grading where this has been accredited. Registration on the scheme is not restricted to JIB graded employees and is therefore open to all electricians who can satisfy the registration criteria.

Front of card

Back of card

*Figure 1.6      JIB registration card*

The purpose of any grading scheme is to divide the workforce into groups based on an individual's skills, qualifications and/or experience. This grading can then be used to determine the wages and the degree of responsibility that may be expected by each person. Progression from one grade to another is possible by improving academic qualifications, industrial training or further industrial experience.

At each stage progression is by qualification and experience. Equivalent grades from other examining bodies are acceptable to allow candidates from different areas and backgrounds the same opportunities for progression.

3

# Contracts of employment

The JIB code of good practice – employment of operatives states;

*"A contract of employment is a voluntary agreement by which an employer and employee will regulate their legal relationship during the course of the employment."*

The fact that it is a voluntary agreement does not mean that it need not be entered into. The implication is that the content of the contract may be voluntarily agreed between the employer and the employee. In practice many companies use a standard contract of employment and clauses may be added to this by mutual agreement.

Most new entrants to the industry come in as trainees. As such they will normally be expected to enter into a Training Contract with their employer. This will frequently be a Modern Apprenticeship Pledge instead of a separate training contract. In this document the trainee and the employer each agree to specific responsibilities under the agreement and circumstances for termination of the contract are listed. Both parties, provided the terms are acceptable, sign the contract and these signatures are witnessed. As the trainee may be under the legal age requirement to enter the contract there is provision for a parent or guardian to sign, indicating their consent for the trainee to be bound by the agreement.

The company is required by law to give the following information regarding conditions of employment:

- holiday entitlement and pay arrangements
- periods of notice
- grievance procedure
- health and safety at work
- sickness entitlement
- pension details

On completion of the training contract the employee should have undertaken the basic training requirements for the industry and is likely to be employed as an electrician. This will involve the signing of a new contract of employment stating the agreed conditions that will exist between the two parties.

A typical contract will normally include fundamental details such as normal working hours, annual holidays and an undertaking on both sides regarding the standard of work.

Any employee who has not entered into a contract of employment has no proof whatsoever of the conditions of his employment with regard to hours, period of notice for dismissal, holiday entitlement, rates of pay, position for which employed, insurance cover and so on. This situation is a distinct disadvantage to both the employee and the employer.

OPERATIVE'S COPY

**ELECTRICAL OPERATIVES (including Apprentices) employed under JIB Terms and Conditions—the constituent parties to the JIB are the Electrical Contractors Association and the AEEU.**

**STATEMENT**

**of particulars of terms of employment pursuant to**

**THE EMPLOYMENT PROTECTION (CONSOLIDATION) ACT, 1978**

**as amended by the Trade Union Reform and Employment Rights Act, 1993**

**and the**

**CONTRACTS OF EMPLOYMENT AND REDUNDANCY PAYMENTS ACT (NORTHERN IRELAND) 1965**

**Amended by Industrial Relations (N.I.) Order 1976**

Please type or use capitals with ball point pen—make sure **all** copies are clear

1. From: Name of Employer ......
   Address ......
2. To: Name of Employee ...... (Surname) (Full Christian Names)
   Address ...... (Number/Name of House) (Street or Road)
   ...... (Town) (County) Postal Code
   National Insurance No. ..../..../..../..../.... Works No. ....../None*
3. Date of commencement of Employment ...... (Day) (Month) (Year)
   *(If previous Employer is associated company or same company under new ownership the date on which the continuous period of employment began should be stated.)*
4. Date of this Statement ...... (Day) (Month) (Year)
5. (i) Position: ...... (Insert JIB Grading) /Apprentice*
   (ii) Current Rate of Pay: £ ......
   (iii) Hours of Work: ......
   (iv) Pay period: *weekly/monthly
   (v) Holiday entitlement:
   (i) Annual Holidays ......
   (ii) Public Holidays ......
   (vi) You have been:
   (a) *Shop Recruited in which capacity you are liable to be transferred to work on any site on which your Employer has a contract. Location of shop if different from 1. above;
   ......
   (b) *Locally Engaged in accordance with National Working Rule 12(f). Site initially engaged.

*Figure 1.7 Part of a typical contract of employment*

# Disputes

We can see that agreements, grading and contracts of employment are all vital to provide clarification of the duties, responsibilities and conditions within the industry. Even with all these formalities completed there are still likely to be disputes between employees and employers. These may vary from minor local disputes, such as extra allowances for a particularly unpleasant job, to a national dispute over pay or conditions. In any event there are laid down procedures that should be followed to ensure a satisfactory settlement. We shall now consider these procedures and follow the sequence of events that could occur.

*Remember*

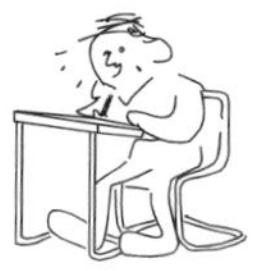

Labour relations involve the following organisations:

AEEU
E.C.A.
JIB

The N.I.C.E.I.C. is the independent consumer safety body.

Agreements, grading schemes and so on are decided nationally by the JIB.

Every employee should have a contract of employment.

There is a recognised grievance procedure operating within the industry.

## Dispute procedure

We will consider a dispute procedure similar to that suggested by the JIB to illustrate a typical sequence of events.

In the first instance the employees should report, to their immediate supervisor, any problems relating to their conditions.

If a satisfactory solution is not forthcoming then the employee's shop steward or staff representative may approach the supervisor concerned, and if necessary the Employer's Representative on site, on behalf of the individual operatives.

In the event of a solution still not being obtained either party may refer to their respective organisations for assistance. If the problem cannot be solved now, it should immediately be referred to the local Regional JIB, in the form of a full written report of the problem addressed to the Secretary of the Board. The Secretary will endeavour to reach a conciliated settlement but failing that he will report to the Chairman of the Regional JIB. An investigation may then be carried out by the regional board which will ascertain the facts and settle the problem within the terms of the JIB rules.

An appeal may be lodged against the decision and, if so, it must be lodged in writing to the secretary of the JIB within twenty-eight days of written confirmation from the regional board. A decision will be made by reference to a meeting of the National Board.

In the broader field of industrial disputes if a settlement is not reached between the parties the dispute may, by joint consent, be taken to an independent body for impartial judgement of the facts and assistance in arriving at a mutually acceptable agreement. The best known of these is the conciliation service ACAS. Fortunately most disputes are relatively minor and are usually dealt with in the first one or two steps of the procedure.

Figure 1.8    ACAS logo

# Part 2

## Contractual relationships

On any site there will be a number of different groups working. These will include:
- the architect or client's representative
- the main contractor
- nominated subcontractors
- subcontractors

and it is a good idea to investigate how these are related professionally during the course of the contract.

**Client:** this is the person who requests the work to be done, the customer as we would call them on small domestic jobs. If the work to be carried out is of any size, which could be from building a single dwelling to a multimillion pound construction, it is likely that the client will engage the services of an architect.

**Architect:** the architect will be appointed by the client in the first instance to interpret the client's wishes with regard to design and construction. In addition to this, the architect may be responsible for the supervision and administration of the contract for the client. It is the architect who acts as the client's agent, negotiator and arbitrator. Any changes to the original requirements, time extensions, quality of work and so on are all under the control of the architect who is, in turn, acting in the interests of the client. In practice the architect will often employ an Electrical Designer to provide specialist information and advice.

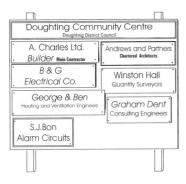

Figure 1.9    Site notice board

**Management Contractor:** on larger jobs it is quite common for the client or architect to engage the services of a management contractor. This company performs the functions of:
- letting main and subcontracts
- programming and monitoring progress of work
- alterations and additions
- standard of construction
- payment of trades

and the like. The management contractor generally takes no active part in the actual construction of the project. In some circumstances a main contractor may take on the role of management contractor, carrying out the above functions in respect of all the other trades involved.

**Main contractor:** this is the contractor who has been awarded the contract to produce the finished work in its entirety. The main contractor is usually a building contractor responsible for the main construction. In order for this to be carried out efficiently specialist contractors are often used. These may be employed to carry out such tasks as wiring, heating and ventilation, alarm and call systems and plumbing. If these trades are not part of the main contracting company, they are employed as subcontractors. A subcontractor may be nominated by the client or the architect.

**Nominated subcontractor:** Because of the specialist nature of some of the work to be carried out the architect or the client may select a particular contractor for this work. The nominated subcontractor's contractual responsibility will still be to the main contractor unless the client issues a separate contract to the subcontractor. There are a number of advantages to being a nominated subcontractor. One advantage is the clause in the contract enabling the client to pay the nominated subcontractor direct, in the event of default by the main contractor. Typically these could be the computer installation contractor or the fire alarm commissioning contractor, who may also be the nominated supplier of the equipment.

**Subcontractor:** These are usually selected by the main contractor who specifies the terms offered for their work. The subcontractor is not recognised by the architect and is not mentioned in the main contract. The work that they carry out is usually that which could be done by the main contractor but which he finds cheaper or more convenient to contract out. These contractors are often involved in work allied to other trades such as fitting insulation to pipework installed by heating engineers.

In addition to those considered above, there may be consulting engineers, who advise on specialist areas of work, and clerks of work whose main function is the inspection of workmanship and materials. Figure 1.10 shows the typical relationship between each group involved in the contract.

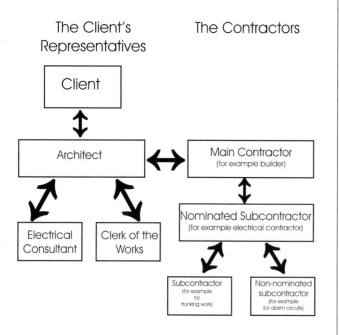

*Figure 1.10*

## The electrical contractor

Most electrical contracts are tendered against a specification drawn up by the electrical consulting engineer, employed by the architect or client. The successful contractor will be referred to as the electrical works contractor. The work is carried out under a separate contract to the main contractor and therefore the status of the electrical contractor will be that of a subcontractor. The effect of this is that the main contractor may be regarded as the customer. The electrical contractor will be bound by the terms of the contract with the main contractor.

## Suppliers

There will also be suppliers of equipment involved. These may be directly appointed by the client or his consultants or selected by the contractor. Specialist suppliers of such items as fire alarm equipment, lighting fittings, public address equipment and standby and UPS (Un-interruptible Power Supply) supplies are often specified. More common equipment may be subject to compliance with a particular standard or group of manufacturers. The client or the consultant should be formally approached to agree any variation to their prescribed suppliers.

It may be possible to use an alternative supplier for any of the required equipment but this will **always** have to be agreed with the client or the consultant. If a supplier has been nominated by the client or the consultant there may already be an agreement between those two parties. A good case for change, based on technical specification or cost saving, will be necessary for a change to be considered.

*Figure 1.11     Supplier's vehicle*

## Advisers

On many contracts, advisers are necessary in order to ensure that all the statutory requirements have been complied with. These include fire officers, district surveyors, building inspectors and the like. Advisers may also be engaged to provide advice on specialised areas of the installation or construction. This may involve the requirement for EMC (Electro-Magnetic Compatibility) and its minimisation, public address and alarm systems, computer cabling and the like. Advisers differ from consultants or contractors in so much as their involvement in the project does not extend to any contractual or management control.

The advisers for statutory requirements have the power to accept or reject the proposals made for compliance. On completion they will also assess the project compliance and should this not meet the requirement it will not be accepted. This may result in considerable remedial work and delay in occupancy of the building. Their comment and advice should therefore be sought throughout the construction process.

Advisers engaged for specialist areas may only be required for one-off discussions or they may be consulted throughout the contract, dependant upon the complexity of their particular expertise. Whilst these advisers can indicate the requirements and suggest methods of achieving the desired result they have no contractual powers and would not be in a position to reject the installation upon its completion. However ignoring their advice could result in an installation which does not meet the client's requirements or satisfy the performance criteria. Should implementation of their recommendations have cost or contractual implications these should be resolved with the client or consultant before they are implemented.

All the parties mentioned so far may have some input into the project at the design, construction and commissioning stages. It is likely that the labour-only subcontractor and the suppliers will have no design involvement. The specialist subcontractor and supplier may, however, have a significant input at the initial design stage which could continue throughout the contract.

## Working relationships

Whilst work is being carried out liaison is needed among all the parties involved in the project, particularly on site. Good industrial relationships are of considerable importance on site to ensure the cooperation of contractors and the smooth running of the job. Any dispute between contractors can result in delays in the site progress, which may result in a time penalty being invoked. The main contractor may be penalised at a high cost per day for failure to complete on time. In such a case it would be logical to apportion blame to any subcontractor whose actions were responsible for this turn of events.

Poor relations between two subcontractors could result in them both being involved in a court case to recover the costs. Providing the main contractor feels that the responsibility can be substantiated, the company's legal representative will have no hesitation in pressing a claim, which could be in the order of thousands of pounds.

*Figure 1.12      Good relationships on site*

In addition to the relationship with other contractors we must also consider relationships with the customer or client. In an industry where competition is fierce a personal recommendation can count for a great deal. This applies to both the small contract, such as for additions to domestic wiring, and dealings with large contractors.

If a contract is carried out quickly, efficiently and with good will and courtesy then the main contractor or client is likely to use the same contractor on future projects. Remember there are a number of advantages to being a nominated subcontractor.

A courteous and civil approach will help in any negotiations that may be needed. All of this will help us to run the job with the minimum of inconvenience. This may bring more work to the company, improving the employment prospects for the future. The employer needs the staff on site to act as the representatives of the company. Staff actions and attitudes will be taken as representative of the company as a whole.

## The client's requirements

Some clients will be aware of the technicalities involved in the work while others will have no knowledge at all. Similarly some will be cooperative while others will be difficult and obstructive.

It is important to establish what it is the client requires and every effort should be made to enable them to realise their intentions. Your clients will generally have at least an outline idea of what they require, some will have quite specific ideas as to what they want. In any event the first thing we need to establish is what these requirements are and how we can ensure that the client gets what they want.

In order to do this we may need to carry out some research into the available products, characteristics and suitability of the supply and other relevant details. Once these have been obtained we need to confirm the acceptability of the proposed equipment and produce a design proposal for the installation. This will need to be discussed with the client and any cost or programme implications for the proposals will need to be provided and discussed with the relevant parties. Once the details have been agreed, we shall need to develop the detailed design for the installation.

*Figure 1.13      Programme changes need to be discussed with the client*

It is important to ensure that the client is advised of changes and developments as they occur in order that the finished product has been agreed and satisfies the client's requirements. This will ensure a minimum of remedial work and "wasted time and material" to achieve the completion of the project.

On larger projects the client may have entrusted the final details to the consultants and architects employed for the project. In such cases many of the details may already have been decided, but inevitably there will be some involvement for the electrical contractor. This may be as little as the co-ordination of the final routes of cables and containment systems with the other services. In some cases it will involve the determination of cable and containment system routes, calculating conductor sizes, construction and capacity of control gear and distribution boards and the like.

It is important to ensure that, irrespective of the size of the project or the extent of the design involvement, the clients or their representatives are consulted, advised and in agreement with the proposals throughout the project.

If your clients appear to have little or no understanding of what the job entails it is important not to make them feel foolish. Explain simply and clearly any parts of the work that are not understood. It may be necessary to ask questions of the customers in order to establish the extent of their understanding.

---

*Remember*

- A courteous approach may result in the customer returning with more work in the future.

- Customers who are not treated with respect will be likely to complain and they may also become aggressive.

- Your employer will expect all staff on site to be representatives of the company whether dealing with clients, the contractor or subcontractors.

---

## The contract

In the world of business a contract is usually in the form of a written legal agreement. This agreement will cover all aspects of the terms and conditions of the job. Both the contractor and the client involved in the contract must agree to all the terms and conditions before the contract is signed.

A contract would, as a minimum, need to include such factors as:

- where the work is to be carried out
- when it will be started and when it is to be finished
- exactly what is included
- what standards it will conform to and
- which wiring system will be used

If a specification was issued then the contract will have been tendered to this specification. The electrician must ensure that the work is carried out to the specification

## Site visitors

It may be necessary for inspectors, advisors, company's or customer's representatives to visit the site on occasions. A record is kept of who is on the site at all times and a procedure to establish the identity and authority of the visitor and the purpose of their visit should be in place.

When visitors are to be allowed on site they must be advised of the Health and Safety Rules which apply to any safety equipment required. The instruction or information must be clear and easy to understand and preferably available in written format, such as an advice poster. Where particular hazards exist it may be necessary to provide an escort to accompany the visitor whilst on site.

Try this
Draw a line diagram showing the relationship between the parties which may be involved between the conception and completion of a project.

## Points to remember

Organisations involved in the electrical industry are:

- The AEEU: a body representing the employees' interests
- The ECA: a body representing the employers' interests
- The JIB: a body under which the above combine to determine the conditions within the industry as a whole
- The NICEIC: the only independent body representing the consumer's interest

Agreements and Grading Schemes are used to determine the working conditions and salaries of operatives.

Contracts of employment between the employer and employee are used to agree the

- hours of work
- holiday entitlement and pay arrangements
- periods of notice
- grievance procedure
- health and safety at work
- sickness entitlement
- pension details

Contractual relationships between clients, consultants and contractors form an important part of the control and operation of a project.

Good working relationships between all those involved in the job are important for the success of the project.

1. List the items that should be included in an employee's contract of employment.

2. List three typical clauses which could be part of an agreement related to conditions of employment.

3. Draw a line diagram to show the contractual relationships between the parties involved in a project.

4. State the steps which should be taken when visitors attend site to ensure their safety.

5. List the items which should be included in a contract to carry out an electrical installation.

# 2

# Health and Safety, Hazards, Regulatory Requirements

Before you start work on this chapter, complete the exercise below to ensure that you remember what you learned earlier.

An electrician would have his terms and conditions detailed in an agreed _____ __ _____.

The Electrotechnical Certification Scheme registers electricians according to their individual _____, _____ and _____.

A contract to carry out an electrical installation may involve the _____, an architect, _____ _____ and main contractor.

On larger projects the electrical contractor will usually be a _____ to the main contractor.

Good _____ _____ are important for the success of any project and it is important to ensure that the client's _____ are acknowledged and achieved.

The _____ between the client and the contractor determines the details of when, where and how _____ will be achieved.

## On completion of this chapter you should be able to:

◆ identify the Acts and Regulations which apply to installation work
◆ distinguish between Statutory and non-Statutory regulations
◆ describe briefly the scope of the Health and Safety at Work etc. Act
◆ identify Regulatory requirements with reference to fire
◆ outline the possible effects of failure to accurately assess and determine Regulatory requirements
◆ state methods of confirming Regulatory and regional planning requirements
◆ state the importance of maintaining accurate results
◆ state how Regulatory and planning requirements may have implications for the electrical contractor with reference to finance, legislation, resources and technical requirements

# Part 1

# Site regulations and codes of practice

Whilst we are engaged in our work on site there are a number of Acts, Regulations and Codes of Practice that we must be aware of. Some of these are statutory, which is to say they are legally enforceable, and failure to comply with these can result in prosecution. With this in mind it is as well for us to make sure that we are aware of these acts and regulations and their implications for us. We shall consider some of the most common of these here but remember it is vital to check precisely what the requirements are for each site.

## Statutory regulations

These regulations are those that are backed by Acts of Parliament and are enforced by law. The lists shown here are those that are of particular interest. First there are a number of Acts and Regulations relating to the construction industry generally and to work in any area.

There are a number of requirements particularly aimed at construction sites and the fundamental requirements are detailed in the Health and Safety Executive publication "Health and Safety in Construction", HS(G)150, available from HSE Books.

The most common of the general Regulations are:
• Health and Safety at Work Etc. Act, 1974
• The Management of Health and Safety at Work Regulations 1999
• The Construction (Health, Safety and Welfare) Regulations 1996
• The Construction ( Head Protection) Regulations 1989
• Lifting Operations & Lifting Regulations 1998
• The Construction (Design and Management) Regulations 1994

- The Provision and Use of Work Equipment Regulations (PUWER) 1998
- The Personal Protective Equipment at Work Regulations 1992
- The Manual Handling Operations Regulations 1992
- Factories Act 1961 and the Regulations made under this act
- Construction (General Provisions) Regulations 1961
- Offices, Shops and Railway Premises Act 1973
- Occupiers Liability Act 1984
- Control of Substances Hazardous to Health Regulations 1999 (COSHH)
- Noise at Work Regulations 1989
- Workplace (Health, Safety & Welfare) Regulations 1992

Whilst this appears to be a formidable list of Acts and Regulations, sound guidance on the application of these requirements is given in HS(G)150 which also refers out to other publications where appropriate. You should have a copy of this guidance at hand when considering the health and safety requirements for your work.

We shall briefly consider the general health and safety requirements in this chapter and the "Health and Safety in Construction", HS(G)150, should be available for reference during this activity.

In addition to the general Regulations there are also those with particular relevance to electrical installations. The most significant of these being:
- Electricity Supply Regulations 1988
- Electricity at Work Regulations 1989
- Cinematograph Regulations 1955, made under the Cinematograph Act 1909, and/or Cinematograph Act 1952
- The Electromagnetic Compatibility Regulations 1992 as amended
- *The Highly Flammable Liquids and Liquefied Petroleum Gases Regulations 1972
- *The Petroleum (Consolidation )Act 1928

*These last two come under the Home Office. Under the Consolidation Act of 1928 the local authorities are empowered

*Figure 2.1     The Petroleum (Consolidation) Act 1928 comes under the Home Office*

to grant licences for storage and the authority may impose such conditions as they think fit.

There are Local Authority requirements for Caravan Sites and the electrical contractor should be aware of these requirements when carrying out work in these locations.

Other legislation relating to particular activities may also contain reference to electrical installations and Appendix Two of BS 7671 provides further details on Statutory Regulations and their relevance to electrical installations.

## The Health and Safety at Work etc. Act 1974

The Health and Safety at Work etc. Act applies to all places of work and to both employers and employees. We all have responsibilities under the act so it is worthwhile for us to consider it in more detail.

So what is it and what does it do?

The Health and Safety at Work etc. Act covers all aspects of health and safety at work. It lays down the formal responsibilities of the employer and employee. Being a Statutory Act these responsibilities are enforced by law and ignorance will not be accepted as a plea in the event of prosecution.

*Figure 2.2     Training in the erection, use and inspection of the tower scaffold should be given to ensure as far as possible safe working conditions.*

The entire content of the Act is too great to reproduce here but we can summarise the scope of the Act as follows:
- maintaining or improving standards of health and safety and welfare of people at work
- protecting others against risks arising out of work activities
- storage and use of dangerous substances
- controlling airborne emissions from certain buildings
- provision of safe plant and conditions of work as far as is reasonably practicable

- provision of information, training and environment to give as practicable as possible safe and healthy working conditions
- employers to consult safety representatives and union representatives and publish their safety policy arrangements
- employers and employees to ensure that their activities do not endanger others
- persons in control of non-domestic premises, used as a place of work, to ensure that the premises themselves and the plant and so on in them do not endanger the people using them.
- designers, manufacturers and importers of goods used at work to ensure that as far as practicable that they are safe in all conditions of installation, use and maintenance
- employees to take reasonable care not to endanger themselves or anyone else with their work activities. Also, to cooperate with their employers in meeting statutory requirements.
- a duty on everyone not to misuse or damage anything provided for health and safety under a statutory agreement.
- employers may not charge their employees for anything done or equipment provided under a statutory agreement for health and safety.

Figure 2.3    *Employees should take reasonable care not to endanger themselves or anyone else with their work activities.*

Under the Health and Safety laws your employer must display a poster which outlines the responsibilities of both yourself and your employer. The Health and Safety Executive publishes a number of leaflets and guides to the health and safety regulations and these may be purchased from Her Majesty's Stationery Office (HMSO).

Figure 2.4

The effect of contravening the requirements of the Health and Safety at Work etc. Act can be a fine of up to £2000 on summary conviction. In the case of an indictment there is no monetary limit. For more serious offences there is a possibility of imprisonment for up to two years and/or a fine. It is important to remember that individual operatives, directors and/or managers are liable to prosecution under the Act and the responsibility is not the "company's".

Rather than parliament having to make all the legislation relating to health and safety it gives the power to the Secretary of State for Employment. This power is exercised through the Health and Safety Commission who draw up the specific regulations and codes of practice on health and safety matters.

The objective of the act is to give a single comprehensive and integrated system of law dealing with the health, safety and welfare of people at work and health and safety of the public whilst any work is being carried out.

The full scope of the act is too large to present here but a copy of the act should be available for your inspection at all places of employment and at public libraries. It would be a good idea to take a look at the scope of the act just to familiarise yourself with the areas of responsibility that affect you.

Every person, employer and employee alike, has duties under the act and these responsibilities cannot be passed off. For example, whilst you are at work you see a manhole from which the cover has been removed and no barrier erected. YOU have a responsibility to report this so that a barrier can be put up and you should not just ignore it because you are not involved directly with the work concerned.

Figure 2.5    *YOU have a responsibility to report this type of incident so that a barrier can be put up.*

"Health and Safety in Construction", HS(G)150, may appear at first glance to be specific to work on new construction projects. However it does contain good sound guidance for all workplaces where electrical installation work is being undertaken. We shall consider the requirements for some of the more relevant general activities, before we move on to the "electrical installation" related ones.

# Protective equipment

Every person engaged in work activities has a responsibility to ensure their own personal safety. This requirement is reflected in the various acts of legislation covering personal protective equipment. These include the legislation covering hard hats, footwear, goggles and safety spectacles, outdoor and high visibility clothing and gloves.

Whenever a work activity involves a risk which can be minimised by the use of protective equipment, the individual involved has a responsibility to ensure that the appropriate equipment is used. Some of the hazards for which the protective equipment may offer protection against are given below.

## Footwear
- heavy materials dropped on the feet,
- prevent penetration by nails and other sharp objects from below (standing on objects)

Figure 2.6        Footwear protection

## Safety goggles/ spectacles
- dust and flying debris, such as when chasing out brickwork
- sparks from cutting operations such as disc cutters
- chemical splashes, such as PVC conduit solvent
- airborne debris such as cement dust on construction projects

Figure 2.7        Eye protection

## Hard hats
- material being dropped by persons working above
- material being kicked into pits, excavations and service wells
- material falling from lifting activities such as hoists and cranes
- dangers from fixed protrusions in the work area, such as mounting brackets and pipe clamps.

Figure 2.8        Head protection

## Outdoor clothing
- the effects of wind and rain on the general health of the individual
- particular effects of the elements on long term exposure to the elements, prolonged outdoor working. This may involve work with little or no physical movement such as the termination of cables in a feeder pillar, or extreme physical activity such as digging.
- protection against some of the elements may be provided by portable shelters, such as the framed work tent used by a cable jointer.

Figure 2.9        Thermal socks

## Gloves

- wherever activities may cause risk of skin penetration such as wood or steel splinters
- where the nature of the material may cause physical skin abrasion such as moving concrete blocks and bricks
- the materials being handled may be hazardous by either being absorbed into the skin or by transfer to the mouth through contact with food

*Figure 2.10    Hand protection*

## High visibility clothing

- where work is carried out in an area with a high risk from "passing traffic" such as working on street furniture, street lights and bollards
- areas with vehicular traffic such as loading bays and docks
- areas where there are lifting operations being undertaken, such as the construction of high rise buildings.

*Figure 2.11    High visibility clothing*

## Ear defenders

- exposure to high noise levels whether over a long or short period of time.

*Figure 2.12    Ear protection*

All of the above provide some indication of the types of hazard which may need to be considered. It is not an exclusive list and there may be other considerations to be made. Further information regarding the use and requirements for protective equipment can be found in the appropriate Health and Safety Guidance and relevant legislation.

# Part 2

## COSHH

The Control of Substances Hazardous to Health (COSHH) Regulations 1994 make it a legal duty to assess substances used in the work environment, determine any health risk involved and prevent exposure or adequately control the risk.

The risks involved are generally in relation to three main categories
- swallowing or eating contaminated materials
- breathing fumes, dust or vapours
- direct contact with the skin or eyes

and an assessment should be made of each substance and appropriate control of the substance detailed.

*Figure 2.13    Concrete dust may be harmful to lungs over the long term*

Manufacturers and suppliers are obliged to detail any information regarding their products and give instructions on the appropriate precautions. Anyone using such products has an obligation to ensure that the appropriate precautions and protective measures are taken.

There are some hazards which may already be present on the site such as sewer gas or land contamination which must be taken into consideration by all operatives. The persons responsible for the site have a duty to ensure that all personnel are aware of the requirements and that suitable precautions are taken.

Further information and guidance can be found in HS(G)150, the COSHH Regulations and other Health and Safety guidance.

There are particular Regulations controlling both asbestos and lead, these being

> The Control of Asbestos at Work Amendment Regulations 1998
> Working with Asbestos Regulations 1993
> and
> The Control of Lead at Work Regulations 1998

These regulations control the use of these products in the working environment and should be considered whenever work involving these materials is to be undertaken.

---

*Remember*
If in doubt ask the site manager and obtain the appropriate guidance. The HSE produces a considerable amount of guidance material. Some material  is free and some at a minimal cost. If you're not sure ask and obtain the HSE recommendations **BEFORE** you commence work.

---

# CDM

The Construction (Design and Management) Regulations 1994 covers the requirements for the management of health and safety on construction projects. We shall not be able to address the requirements of the CDM Regulations here but there is an approved code of practice, produced by the HSE, which explains the requirements and provides guidance on the implementation.

The principal is that those involved in the construction project must consider the safety requirements, appropriate to the construction project, and ensure that appropriate measures are taken. The extent of the required measures will vary dependant upon the type, size and scale of the project concerned.

# Reporting accidents and work related diseases

The Reporting of Injuries, Diseases and Dangerous Occurrences Regulations 1995 (RIDDOR) require that certain accidents which happen on site have to be reported to the Health and Safety Executive.

This requires the notification without delay, usually by telephone, of serious and fatal accidents. This action must be followed by the presentation to the HSE of a completed accident form (F2508) within ten days. A similar requirement exists for dangerous occurrences such as the failure of a crane or lifting device or contact with overhead lines and collapse of

a building or scaffold. Other less serious incidents must be notified to the HSE using the F2508 form within ten days of the incident.

Incidents of specific work related diseases must also be notified to the HSE by the use of an F2508A form.

---

*Try this*
List the details regarding the health and safety legislation available at your place of work, including posters and notices.

---

# Accident book

Wherever people are employed under the provision of the Factories Act, The Offices, Shops and Railway Premises Act or where ten or more people are normally employed then the employer is responsible for keeping an accident book.

Every accident at work, even the minor ones, should be recorded in the book and there are certain facts that must be entered.

An accident form should contain the following information:
- date and time of the accident
- place where the accident took place
- brief description of the circumstances
- name of the person injured
- sex of the person injured
- age of the person injured
- occupation of the person injured
- nature of the injury sustained

Serious injuries are referred to as major injuries and these are defined as

- fracture of the skull, spine or pelvis
- fracture of any bone: in the arm, other than the wrist or hand, in the leg, other than in the ankle or foot
- amputation of a hand or foot
- loss of sight of any eye
- any other injury which results in the injured person being admitted to hospital as an inpatient for more than 24 hours

In addition every accident to an employee which results in the inability to work for more than three consecutive days must also be reported.

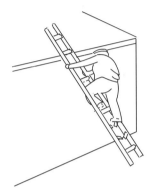

*Figure 2.14      An accident waiting to happen!*

---

### Dabson Electrical Co.

## Accident Report Form

This form must be completed in the event of any accident/injury occurring at the above premises or whilst working for the above company.

## Date and time of accident:
*Date:*                              *Time:*

## Place where the accident took place:

## Brief description of the circumstances:

## Injured person:
*Name:*

*Sex:*

*Age:*

*Job title:*

*Nature of injury sustained:*

## Your details:
*Name:*

*Job title:*

*Figure 2.15      Typical accident report form*

Whilst you are involved in installing and terminating an SWA cable one of your electricians cuts himself, quite deeply, with a knife.

Write, in your own words, an accident report that contains all the information required under the Health and Safety at Work requirements.

# Building regulations and building control services

The Building Regulations are approved by Government and detail the minimum standards of building work for the construction of commercial, industrial and domestic buildings. They define what is regarded as building work and procedures for ensuring that the work meets the minimum standards.

Building Regulations will normally apply to work involving

- construction of a new building or an extension to an existing building

- installation, extension or alteration to drainage

- structural alteration of an existing building

- change of use of an existing building

- installation of a heat producing appliance (gas appliances installed by people approved under the Gas Safety Regulations are usually exempt)

- installation of unvented hot water storage systems

- installation of cavity wall insulation

Building Regulations will not normally apply to work involving
- carrying out minor repairs

- installation of new sanitary fittings

- replacement of roof coverings with identical material

The above lists are not finite and advice should always be sought from the Local Authority concerned.

The installation and replacement of electrical wiring does not currently come under the requirements of the Building Regulations; however, this situation is currently under review.

The Building Regulations include a list of requirements which are intended to safeguard the health and safety of people in and around buildings. They also provide requirements for the provision of access for disabled persons and energy conservation.

The Regulations comprise a number of Approved Documents which detail the requirements for aspects of the construction. These provide detail of the requirements and information as to how these requirements may be achieved. Approved Document B for example considers the requirements for Fire Safety including the installation of smoke detectors and the like in domestic installations.

*Figure 2.16      Smoke detector*

# Building control

The Local Authority has a general duty to ensure that building work complies with the Building Regulations. This is normally carried out by the Authority's own Building Control Surveyors who are employed by the Local Authority Building Control section. These are often referred to as the "DS" or District Surveyors. They will check and approve plans, and will regularly inspect the work during the construction and upon completion to ensure compliance with the Building Regulations.

Alternatively the building control may be carried out by an Approved Inspector who will also check and approve plans and will regularly inspect the work during the construction and upon completion to ensure compliance with the Building Regulations.

# Non statutory requirements

There are also a number of non statutory codes of practice and regulations that apply to electrical installation work. Of those directly concerned with installations the best known are:

BS7671: 1992 (the IEE Wiring Regulations)

The British Standards Institute Codes of Practice

The British Standards Specifications

These are non statutory regulations which mean that they have no legal enforcement in their own right. Whilst these are advisory, they may be included in terms of contract, making them part of that contract. Compliance can then be legally enforced but only under the laws of contract.

Whilst BS7671:1992 is non statutory, by complying with the regulations we can comply with a number of statutory regulations. Complying with any one of the statutory regulations does not, however, mean that we will comply with BS7671:1992. Within the industry it is the accepted practice to comply with the requirements of BS7671:1992 and thus achieve compliance with most of the standard statutory requirements. Specific regulations should **ALWAYS** be checked to ensure full compliance is achieved.

These non statutory requirements are recommendations of good practice for both the installer and the designer intended to ensure safety to all users of electrical installations.

**Note:** In Scotland BS7671:1992 is a part of the Building Regulations which are statutory and so in Scotland BS7671:1992 is enforceable by law.

*Try this*
You are the electrician in charge of a contract to install electrical services in a new factory. List the statutory acts and regulations that will apply whilst carrying out the contract.

*Remember*

There are a number of statutory and non statutory requirements that apply to electrical installations specifically.

Health and Safety laws apply to all employers and employees.

Both the employer and employee have defined responsibilities under the law.

Individuals may be liable to prosecution for disregarding their responsibilities.

Ignorance of these requirements is no excuse for failing to comply.

# Part 3

## The implications of regulation

We must now consider the implications of the statutory and non statutory regulations and codes of practice upon our electrical installation. The requirements of these regulations must be considered from the design stage through the construction, commissioning and the energising and placing into service of the finished installation. Some of these requirements will have an effect on the design of the installation itself and others on the method of installation and activities carried out by the operatives actually doing the work.

### Design

First we shall consider the implications for the design of the installation itself. The need to ensure the design proposals meet the statutory requirements is paramount and, in general terms, the standards will be the Electricity Supply Regulations 1989, the Electricity at Work Regulations 1998 and the Building Regulations. As we discussed earlier, the use of BS 7671, whilst a non-statutory requirement, does provide a means of achieving compliance with the statutory requirements.

Failure to comply with the requirements of the statutory regulations can result in prosecution, usually through a government body such as the local authority or the Health and Safety Executive, in connection with the health and safety implications of the design non compliance. Inevitably the failure to comply with the statutory requirements will involve remedial action to remedy the situation, which will undoubtedly involve some additional cost. In respect of the design of our installation this may be quite considerable as the basis upon which the installation is founded would be flawed. These prosecutions would be under Criminal Law as the

statutory regulations are passed by parliament and compliance is a legal requirement.

The necessary remedial work may also cause disruption to the building programme and inconvenience to the building user which could involve loss of revenue and damage to the finish. As a result, there may be a legal action from the client in regard to the failure to meet the legal requirements for the installation and further action to recover any loss incurred as a result.

The cost involved in the recovery of associated loss may be quite considerable, particularly in connection with penalty clauses and loss of production or revenue. Dependant upon the project there may also be a claim in respect of the loss of "goodwill and business opportunity" due to the late opening or commencement of business as a result.

Failure to meet the non-statutory regulations will generally result in an action being brought by the client, as the installation provided may be considered as unsuitable.

In addition we need to consider any particular regional requirements as these must also be incorporated in the design and construction of the installation. The various electricity supply companies have their own particular requirements with regard to PME supplies (TN-C-S systems). Additionally there may be regional requirements for the installation of fire detection and emergency lighting systems, licensed premises, houses of multiple occupation and the like.

It is important that these "regional" requirements are considered at the design and construction stages. Failure to do so is likely to result in the installation not being accepted and as a result remedial action will be required.

*Figure 2.17*     *Working on a construction site*

### Construction

The construction work must not only meet the requirements of the design but also be carried out in accordance with the health and safety requirements. Failure to do so may result in action being taken by the Health and Safety Executive for the improvement of the site. This could involve additional safety measures and cause delay to the progress of the work, again involving additional costs.

Should a notifiable incident occur during the construction of the installation there is usually an investigation by the HSE. If, during the installation, statutory health and safety regulations have been breached then a prosecution by the HSE is likely to follow. The investigation could result in a delay to the progress of the work and involve additional cost for that delay. It may also attract a fine and legal implications as a result of the incident.

## Confirmation of requirements

In order to ensure that our installation complies with the regional and regulatory planning requirements, it is important that the relevant authorities and regulators are consulted. By reference to the recognised standards we are able to establish the fundamental requirements. The consultation should consider any variation or additions peculiar to the region.

As we establish the requirements, it is important that we keep records of the agreed criteria and confirmation of any proposals. This should be undertaken at the planning and design stage of the work and any requirements for inspection during the work process should be established. The relevant inspectors should be notified in good time, to ensure the construction process is not delayed and records of the visits should be maintained. Any additional requirements or modifications should be recorded and, where these affect the programme or cost, they should be raised with the client or professional team.

Any testing carried out should be recorded, and accurate and detailed records of the results maintained. This serves not only to demonstrate that the necessary precautions and tests have been undertaken, but also the condition of the installation at the date when the tests were carried out.

*Figure 2.18    The District Surveyor arrives on site*

## Implications

As we can see the regulatory and planning requirements have implications across all the aspects of our work. At this stage we need to consider, in brief, the effect of these requirements on the following aspects:

**Finance:**
The regulatory and planning requirements may have considerable financial implications for our work. The need to comply with the constraints or additional requirements can place a considerable financial burden on the contractor. The extent of this burden needs to be assessed, at the planning stage, in order that the appropriate costs may be included in the quotation. It is also important to advise the client what additional contingencies have been included at the time of presentation. Any additional changes or modifications should be advised to the client and a suitable agreement reached as to the additional cost.

*Remember*
Failing to consider the regulatory and planning requirements at the time of quotation may result in the full cost of their implementation having to be born by the contractor concerned.
In the case of the electrical installation that would be your company.

**Legislation:**
The design, construction and operation of electrical installations are covered in part by legal requirements which must be observed. Failure to take account of and comply with the relevant legislation will leave the company, and often individuals within that company, liable to prosecution. We all have an obligation to consider the safety of ourselves and those working with us. This extends to those persons using the electrical installation which we have constructed.

*Remember*
Ignorance of the requirements is no excuse in law. Knowing the requirements and failing to comply with them cannot be defended.
Deliberately ignoring the legal requirements will almost certainly result in legal action and may be the cause of real danger to yourself and/or others resulting in death.

**Resources:**
In addition to the direct cost involved in meeting the regulatory and planning requirements, there will also be a level of resource necessary to ensure compliance. This resource will have an impact on the financial aspect of the contract. It may also affect staffing levels, accommodation and the progress of the work. All of these need to be taken into consideration at the planning stage and care must be taken to ensure that adequate

resource is made available, in order that the progress and standard of the work are not impaired.

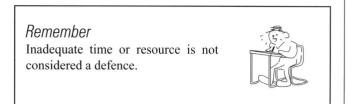
**Technical requirements:**
The effect of regulatory and planning requirements may be to change the technical requirements for the installation. This could result in an additional design burden and additional cost. It is therefore important that the requirements are established at an early stage, in order that both the contractor and the client are aware of the effect on the cost and progress, as well as any performance criteria which need to be achieved. This may result in the use of equipment of a higher specification than that required by the client, in order to meet the regulatory and planning requirements. These need to be advised at the design stage as they may well have a wide-ranging effect on the electrical installation. For example energy requirements, and on the construction due to the physical requirements of plant, equipment and services.

# Working practices

There are many activities undertaken on site which have a direct impact on health and safety. These apply to the individual, those working with and around them and those under their supervision. We shall consider those which have a direct impact on our electrical installation practices.

## Permits to work

Permits to work are used in a large number of instances, many of which have no electrical installation relevance. The purpose of a permit to work scheme is to ensure that activities in areas of risk are controlled and monitored to minimise danger. The risk may be from any source and in particular where the action of others may have a direct influence on the safety of those carrying out the work.

Work carried out in electrical switchrooms, on mechanical plant and processing equipment, aircraft runways, public access areas and secure locations are all likely to be subject to permits to work.

The permit to work system is under the control of a competent person who is charged with the operation of all activities requiring such permits. In order to obtain a permit to work the precise nature of the work involved has to be detailed and a method statement will be required. The purpose of these details is to ensure that the activity can be assessed, the implications of the action considered and an evaluation of the time made. Other actions being carried out may have implications for the proposed work and ultimately affect both the timing and the extent of the work which can be carried out.

*Figure 2.19    Activities in areas of risk are controlled and monitored to minimise danger*

The permit to work identifies the precise nature of the work to be carried out; any other work which is later found to be necessary should be the subject of a further permit. In certain instances, where work needs to be carried out on, say, a main distribution board, it would involve the authorised person to ensure the appropriate distribution board is isolated and made safe. When this is completed the **ONLY** distribution board which will be isolated and safe to work on is the board designated on the permit to work.

For work carried on mains supply networks it is a requirement that, when requesting a permit to work, the network switching details are provided. These are to demonstrate the safe isolation of the part of the network on which work is to take place. Not only must they detail the switching arrangements but also the sequence of switching to ensure the minimum loss of supply. By carefully sequenced switching it is generally possible to isolate a part of the system for maintenance without the loss of supply to the users. Where this is not possible, as in the case of failure or damage, the disruption to consumers should be kept to a minimum. This requirement is contained within the statutory regulations concerning the electricity supply.

The permit to work system is implemented to safeguard everyone involved in the work process and ensure that the statutory requirements are complied with. Ignoring the requirement or exceeding the scope of the permit may endanger yourself or others.

## Method statement

Having mentioned method statements, it is appropriate to consider the use and content at this time. The purpose of the method statement is to detail precisely what is to be done, how it is to be done, any special requirements or actions and the time anticipated for the work to be carried out. Many companies have a prescribed format for producing a method statement but in general they all contain the above details.

Let us consider a relatively simple activity to establish the requirements. It is intended to carry out some maintenance to the high bay lighting in the loading section of a factory dispatch area. This area is subject to heavy vehicle traffic and production schedules. As access equipment is to be located in the main traffic area and requires the isolation of the lighting in a main work area, a method statement and permit to work is required.

In the case of the high bay lighting maintenance the actions detailed on the method statement could be considered as:

- close access doors to area one in the north loading bay at 07.00
- erect mobile scaffold tower at north end of the Bay 1 at 07.15
- install temporary Tungsten halogen portable lighting to operate from loading vehicle power sockets on Bay 1 at 07.15.
- access to switchroom B7 to isolate and lock off circuits B7/DB2L/R12, B7/DB2L/Y12 and B7/DB2L/B12, at 08.00.
- carry out cleaning and lamp replacement to high bay lighting in Bay 1 at 08.30
- access to switchroom B7 to test and energise circuits B7/DB2L/R12, B7/DB2L/Y12 and B7/DB2L/B12 at 10.00.
- test Bay 1 lighting at 10.25
- move mobile scaffold to Bay 2 at 10.50

and so on to allow completion of the work. Where B7 is the switchroom designation, DB2L is the designation given to the appropriate lighting distribution board and R12, Y12 and B12 are the three circuits (Red phase circuit 12, Yellow phase circuit 12 and Blue phase circuit 12).

This enables the person responsible to determine the extent of the activity and the duration. This can then be checked against the programme for dispatch and a date for the work can be agreed. It may be that the method of work needs to be altered, for example the loading bay may be available for a period of say 06.30 until 08.45 each day and the maintenance needs to be carried out at that time, or between the hours of midnight and 06.00.

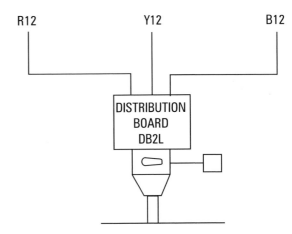

*Figure 2.20      Distribution with circuits R12, Y12 and B12*

# Part 4

## Isolation

Having considered the permit to work requirements we now need to look at one of the most important aspects of the electrical activities undertaken; the isolation of equipment. In order to work safely on electrical equipment, which has been energised and put into use, it is essential to ensure the equipment has been isolated before work commences.

When carrying out isolation we are making an installation, or part of the installation, safe in order that work can be undertaken without danger. The need for safe isolation is a requirement of the Electricity At Work Regulations 1989 and failure to ensure that appropriate isolation is achieved could result in prosecution in the event of an incident.

This responsibility applies whether you are carrying out the work or carrying out the isolation in order for others to work safely. We should

- correctly identify the installation, or part of the installation which needs to be isolated
- isolate that installation, or part of the installation
- confirm the installation, or part of the installation, is isolated
- make sure that the installation, or part of the installation, cannot be unintentionally energised
- ensure that suitable warning notices are positioned to identify the installation, or part of the installation, that is being worked on.
- inform those carrying out the work of the extent of the isolation provided.

Once we have completed the above items, we have fulfilled the duties placed upon us by the statutory requirements. That is we have ensured the installation, or part of the installation, is safe to work on.

## Isolation procedures

When carrying out the isolation of an installation, or part of the installation, there are certain actions which we must follow in order to ensure that isolation is safely achieved. In this case we shall consider the isolation of a lighting circuit in order that it may be extended.

*Remember:*
Isolation is not the operation of a device it is a complete process which must be undertaken if safe isolation and working are to be achieved.

In order to establish whether isolation has been achieved it is appropriate to use a voltage test instrument. HSE Guidance Note GS 38 gives guidance on the safe use of test equipment and you should make reference to this document when selecting and using test equipment. There are a number of units available for testing voltage and the type used is down to personal preference, providing they meet the appropriate standard as indicated in GS38.

The basic procedure for isolation requires the following steps to be taken:

- identify the circuit to be isolated
- check the voltage indicating instrument against a known supply or proving unit
- check the circuit is currently live
- isolate the circuit, using the isolation device
- test the circuit to ensure it is dead, using the voltage indicating instrument
- check the voltage indicating instrument against a known supply or proving unit

- secure the isolating device against accidental or unintentional closure

Providing these steps are followed the circuit can be made safe for work to commence.

Let us consider the purpose of each step in turn.

### Identify the circuit

This may be accomplished by the use of circuit charts, operation and maintenance documents, or the details contained at the distribution board. There are devices available which may be connected to an energised circuit, typically at a socket outlet, which injects a signal onto the circuit conductors. A sensing device, which emits an audible and visual warning, is then used to identify the protective device that supplies the circuit to which the transmitter is connected. In any event it is important to accurately identify the correct circuit to minimise disruption and time wasted.

### Check the voltage test instrument

It is important to test the voltage test instrument we are to use before we begin to ensure that the test instrument is functioning correctly. If the instrument has failed it will indicate the circuit is dead irrespective of its true state. This check may be made using a proprietary proving unit or against a known live supply. In any event it is not safe to proceed until this check has been completed and the instrument has been proved to work correctly.

### Check the circuit is live

It is important to check that the circuit is live before we commence the isolation process. If another action has resulted in the supply to the circuit being isolated then we cannot be sure the circuit is safe to work on. For example, a key/coin operated meter has been installed on a domestic type installation and unbeknown to us there is currently no credit available. We could check that, say the upstairs lighting circuit, is isolated having switched off the circuit breaker. If however the breaker is incorrectly labelled and we have switched off the downstairs lights this would not be immediately apparent. If we were to commence work, and the user of the installation was to refresh the supply, the circuit on which we were working would become live. In order to prevent this it is advisable, wherever possible, to ensure the circuit is live **BEFORE** we begin the isolation procedure. We can then be assured that the circuit has been isolated by the operation of the device we have used and then secured on the off position.

### Operate the isolating device

Use the most appropriate isolating device for the installation or part of the installation to be isolated. In the case of our lighting circuit it would be appropriate to isolate using the designated circuit breaker. Where an "off load" isolating device is to be used the circuit should have the load disconnected before the device is operated.

### Test the circuit is dead

Using the proven voltage testing instrument confirm that the intended circuit has been isolated. If the circuit is found not to be "dead", begin from step one again checking that the appropriate circuit and device are identified.

**Check the voltage test instrument again**
The reason for this check is to ensure that the instrument is still functioning. Should the instrument have failed during the test process then it may be indicating the circuit is dead when, in fact, it is still energised. If the instrument is still functioning then, having shown the circuit to be dead, the circuit may be considered safe to be worked on.

**Secure the means of isolation**
Having confirmed the circuit has been isolated the device used to achieve the isolation should be secured in the off position. Notices should be placed to warn that there is work in progress and that no circuits should be energised without consulting those carrying out the work.

Providing the above procedures are followed then we can achieve safe isolation and comply with the requirements of the Electricity at Work Regulations 1989.

---

*Remember*

Identify the circuit(s) to be isolated

Check the voltage test instrument

Check they are live

Isolate the circuit

Test the circuit is dead

Check the voltage test instrument again

Secure the isolation

---

*Points to remember* ◀ – – – – – – – – – – –

Ensure the work activity and installation comply with the applicable Statutory Requirements.

Be aware of the requirements of the health and safety regulations and your responsibility to meet those requirements at all times.

Use the appropriate safety and protective equipment for each work activity. Where safety equipment is provided by your employer always use it and where an activity requires additional safety equipment do not commence work without it.

Remember to report accidents and work related diseases within the required time scale. Make sure that all the relevant details are recorded without delay.

Consult the non-statutory requirements, regulations and guidance and remember that compliance with the requirements of BS7671 will generally ensure compliance with the statutory requirements related to the electrical installation.

Consult the relevant authorities with regard to additional requirements for the work carried out.

The implications of not complying with the requirements can be serious for all parties involved.

Always use safe working practices and implement suitable procedures for the safety of yourself and others. Follow the requirements of permits to work and ensure that the extent of the permits are fully understood.

Always use the correct procedure for isolation of circuits and installations to ensure that they are safe to work on.

*Self-assessment short answer questions*
1. List the statutory regulations which would apply to both the construction and completed electrical installation within a new industrial unit.

2.  The implications of the Planning and Regulatory requirements may affect the electrical contractor's activities and contract. List the main areas of the contractor's activities which may be affected.

3.  State two effects of failing to adequately assess the requirements of regional and regulatory requirements.

4.  Produce a method statement for carrying out the safe isolation of the circuits involved for the removal and replacement of modular light fittings in an office environment.

5.  List the safety advantages from implementing a Permit to Work System and state one electrical installation activity which would benefit from such a system.

# 3

# Site Administration

Before you start work on this chapter, complete the exercise below to ensure that you remember what you learned earlier.

The electrical installation work carried out is subject to _____ and _____ Regulations covering the safety of the work and the finished product.

Employers and _____ have responsibilities under the health and safety legislation and which cannot be passed on to another or_____.

Appropriate safety and protective equipment should always be used and _____ or incidents reported and recorded.

Failure to consider the requirements of the statutory and regional requirements at each stage of the project can result in _____ and considerable additional expense and may result in prosecution by the _____ ___ _____ _____.

Safe working practices should be used at all times and may involve _____ to work, _____ statements and safe isolation procedures.

---

### On completion of this chapter you should be able to:

◆ state the need for site documentation
◆ fill in site documentation
◆ prepare requisitions for materials from drawings
◆ use a specification to select special tool requirements
◆ read a bar chart to establish the sequence of work for given situations
◆ recognise and draw BSEN 60617 location and circuit diagram symbols
◆ identify and prepare block, circuit and wiring diagrams
◆ interpret designers' drawings

# Part 1

# Documentation

During our everyday working we become involved with some form of administration and record keeping. This generally involves paperwork of some kind for whilst electronic storage and retrieval is becoming quite commonplace, the initial recording of information on site is generally on paper. These documents may vary from the simple wholesaler's receipt for goods bought over the counter, to the work records, ordering, and receipt of materials for a large contract. In addition there will be the need to maintain records and details on the site services, including design, installation and commissioning of the electrical installation and services.

Every piece of information is important and an appropriate record of each must be maintained. In order to simplify these requirements we shall consider a paper record system; other means of keeping records will follow a similar pattern, only the storage and input process will vary.

Each form of record has a specific function so we shall look at the most common, establish their use, and what we should do with them. We shall start with those used for the control of the actual work activity.

### Time sheets

The main function of the time sheet is to provide the employer with a written statement of the activities of staff. You may be only one of a number of electricians, apprentices and technicians working for the same employer. In order for him to pay you the correct wage for the hours you work and to charge for them against the right job you must supply the necessary details. So there will be a minimum amount of information that your employer will require. Whilst this may vary from company to company the basic requirements are as follows:

Name of employee

Week ending

Day of week and date

Name and address of job

Number of hours worked

Travelling time

Expenses/miles travelled

Any allowances

This is not an exhaustive list and some companies may require less information, whilst others may require more. Figure 3.1 shows a typical time sheet layout. These are generally filled out on a weekly basis and are normally countersigned by the electrical foreman, before being sent into the office for processing.

## Day work sheets

These are similar to the time sheet only filled in on a daily basis. They are generally used to cover extra work that has to be charged for or small jobs where the client pays for the work on a time and material basis. It is therefore common to include the amount of materials that have been used in addition to the information required above. The company uses the day work sheet to cost the job for both time and materials and charge the customer accordingly. It is important therefore that the information regarding materials is accurate and the names of all employees and their hours worked are included.

*Remember*
The day work sheet can be used to cover extra work, outside the scope of the original work, on a large contract. They may also be used for small jobbing works such as minor domestic repairs and extra sockets. Figure 3.2 shows a typical day work sheet.

**DOUGHTON BROS.**
Electrical Contractors
### TIME SHEET

Name _____

Week ending _____

|  | Job No. | Time Started | Time Finished | Total for day | Travelling Time | Mileage and fares |
|---|---|---|---|---|---|---|
| Sun |  |  |  |  |  |  |
| Mon |  |  |  |  |  |  |
| Tues |  |  |  |  |  |  |
| Wed |  |  |  |  |  |  |
| Thurs |  |  |  |  |  |  |
| Fri |  |  |  |  |  |  |
| Sat |  |  |  |  |  |  |
|  | Total |  |  |  |  |  |

Operative's signature _____ Date _____

Foreman's signature _____ Date _____

*Figure 3.1      Time sheet*

**DOUGHTON BROS.**
Electrical Contractors
### DAYWORK SHEET

Customer _____

Job No. _____

| Date | No. of men | Time Started | Time Finished | Total | Exp. | Notes |
|---|---|---|---|---|---|---|
|  |  |  |  |  |  |  |
|  |  |  |  |  |  |  |
|  |  |  |  |  |  |  |
|  |  |  |  |  |  |  |
|  |  |  |  |  |  |  |
|  |  |  |  |  |  |  |
|  |  |  |  |  |  |  |

Materials

| Qu. | Cat. No. | Description | For office use |
|---|---|---|---|
|  |  |  |  |

Customer's signature _____ Date _____

Operative's signature _____ Date _____

Foreman's signature _____ Date _____

*Figure 3.2      Daywork sheet*

*Try this*

Assume that you are the electrician in charge of a site with one other electrician. Make out a time sheet to cover his work for a forty-hour week.

Make out a day work sheet for an extra job installing 2 × 13 A sockets in steel conduit. This job will occupy around six hours.

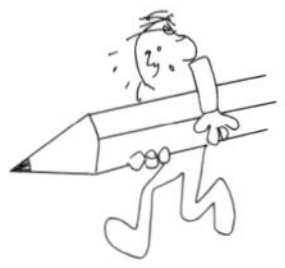

## Job sheets

The Job Sheet normally contains the information necessary for the electrician to carry out the work. It will detail:

The work that is to be carried out
The address
Customer's name
The date on which the work is to be done
Any special instructions such as collecting keys
Any special conditions that exist or equipment in use
It may include details of the materials to be used and type of accessories.

In fact we could regard it as a mini specification and as such it is not practical to use it for larger jobs. It is often used to cover work outside the scope of the contract, such as variation orders, and for small domestic and maintenance work. A job sheet for an alteration on a larger contract may not follow any set layout. These are often written on a blank Variation Order pad by the Clerk of Works, although some companies produce standard layout forms for this purpose. Providing that the information, as detailed in the list above, and the signature of the person authorised to request the work is included then the job sheet is usually valid. Figure 3.3 shows a typical example of a job sheet with the necessary details.

DOUGHTON BROS.
Electrical Contractors
JOB SHEET

Customer ______
Address ______

Work to be carried out

Date to carry out work ______

Special conditions

Special instructions

Name ______
Signature ______
Date ______

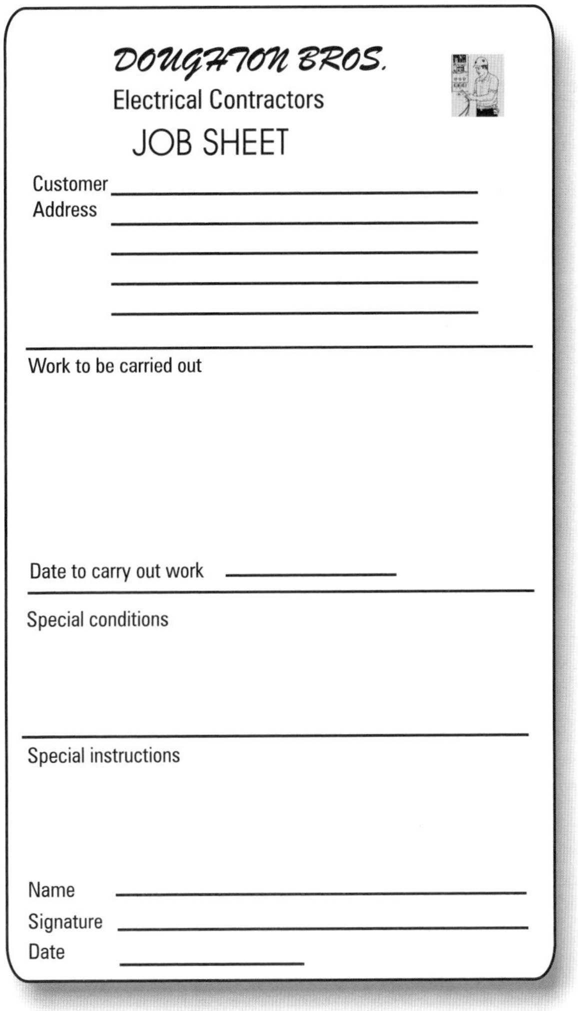

*Figure 3.3 Job sheet*

## Delivery records

These cover the delivery of materials and equipment to the site. There are a number of documents which make up the records for ordering and receiving materials. For us to make sure that we receive the correct documentation and take the right action we must be aware of all the stages involved.

When any material is required the company first sends a written order to the supplier. There are occasions when an order may be placed by telephone and the written part is referred to as a "confirmation order". The order will include the type, quantity, manufacturer of the material required and such details as delivery date and location. The company will retain a copy of the original order. In some instances you may have to raise an order for materials direct from site as a result of a variation that needs immediate action.

Materials delivered directly to the site will be accompanied by a delivery note. As the company representative on site this is the document that you are most likely to have dealings with.

A delivery note should state exactly what materials are being delivered to the site in that particular load. The delivery may not contain everything on the original order placed by the company. This is often because not all the material is required at the same time in order to minimise damage or loss and reduce storage.

The delivery note will state:

- The name of the supplier
- To whom the goods are to be delivered
- The type, description and number of items to be delivered
- A statement as to the condition of the order
- A space for the recipient to sign for the delivered goods
- A statement as to the time period allowed for claims for damaged goods

The "status" of the order could cover a number of conditions and as you will be signing for the receipt of goods it will be an advantage to be aware of these:

- incomplete order
- part of order
- completion order

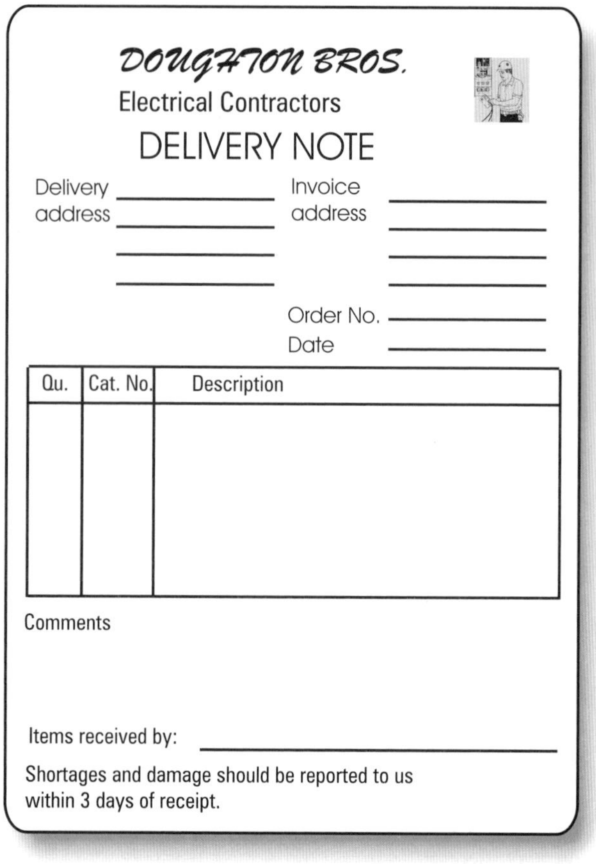

DOUGHTON BROS.
Electrical Contractors
DELIVERY NOTE

Delivery address ____________ Invoice address ____________

Order No. ____________
Date ____________

| Qu. | Cat. No. | Description |
|---|---|---|
| | | |

Comments

Items received by: ____________

Shortages and damage should be reported to us within 3 days of receipt.

*Figure 3.4* *Delivery note*

## Incomplete order

This means that the delivery does not contain all the items that were on the original order. This may be due to the need to have materials delivered at different times, or because some of the order will come direct from the manufacturer. The most likely possibility is that the supplier is out of stock and that the order will be completed as soon as the supplier takes delivery of more stock. As far as we are concerned it informs us that there is another delivery of goods to come. The shortfall in the delivery over what we expected is generally due to one of the above reasons and not as a result of forgetting to order in the first place.

## Part of order

This is very similar to the above. The main difference being that in this case it is most probable that the order is incomplete due to a prearranged cause rather than just being out of stock.

## Completion order

This is the delivery that will finalise either of the previous two provisions. It indicates that the original order placed will have been completely filled on the receipt of these goods.

If the order is completed in one delivery then the above will not apply although some suppliers will stamp the order as complete.

So what is the procedure when goods are delivered on site?

The first job is to ensure that they are unloaded and stored correctly. Whilst the goods are being unloaded they should be checked off against the delivery note. At this stage we are only concerned with the quantities of goods and any obvious damage. Providing the goods delivered match those on the delivery note then you may sign for the items received.

If there are any items missing from the load, and which appear on the delivery note, these should be recorded on the delivery note and signed by both yourself and the delivery driver. It may be an idea to notify the supplier by telephone in such cases to speed up the process of locating and delivering the missing goods.

On the delivery note there will be a statement to the effect that goods damaged in transit must be notified to the supplier within a set time period, often three days. It is in our best interests to check the items delivered within that time and notify the supplier of any damage.

We must file the delivery note and keep it until such times as it is required for reference either by us or the company. Once the goods have been delivered, the supplier will send an invoice to the company. This is really a request for payment for the goods supplied. At this stage the company may wish to check the delivery notes against the invoice received to ensure that they have only been charged for goods which have been received.

It is quite common, especially with suppliers of specialist equipment, to send the invoice along with the goods. In this case the invoice must be sent to the company office straight away. Many suppliers offer a discount for prompt payment and so having the invoice on site, or even lost, can result in loss of discount.

## Reports

It is advisable for the site foreman or engineer to make frequent reports on the progress of the site. This is important because it enables claims for payment to be easily substantiated and made on time. In addition to this it means that any problems that we encounter during the course of the contract can be detected as soon as possible and action taken to correct the situation. Remember that delays can prove to be costly.

In addition to this you will be expected to attend site meetings with the main contractor and discuss any problems that may arise. It is often at these meetings that variations in the work and the orders covering them are discussed. As the company representative you will be sorting out these problems and it is important that any which are not resolved are notified to your main office immediately.

*Try this*

A quantity of goods arrive on site for your use. List the sequence you will follow to receive these goods and place them in store paying particular attention to the delivery note, its content and your dealings with it.

# Part 2

Having considered the receipt of materials on site we shall now consider the records relevant to the progress and work activities.

## Bar charts

| Operation description | Month no. | | | | | | | | |
|---|---|---|---|---|---|---|---|---|---|
| | 1 | 2 | 3 | 4 | 5 | 6 | 7 | 8 | 9 |
| Excavation/concrete | ▬ | ▬ | | | | | | | |
| Brickwork | | ▬ | ▬ | ▬ | | | ▬ | ▬ | |
| Carpentry/joinery | | | ▬ | ▬ | ▬ | | | ▬ | |
| Roof tiling | | | | ▬ | ▬ | | | | |
| Plastering | | | | | | ▬ | ▬ | ▬ | |
| Plumbing | | | | | ▬ | ▬ | | ▬ | |
| Electrical | | | | | ▬ | ▬ | | ▬ | |
| Decorating | | | | | | | | ▬ | ▬ |

*Figure 3.5      Bar chart*

One method used to keep an eye on the rate at which the contract is progressing and compare it with the intended rate is by the use of a bar chart. This is a pictorial method of showing which trade should be doing what, when and for how long. An example of a simple bar chart is shown in Figure 3.5.

The bar chart is a method of showing the intended rate of progress of any activity that we need to monitor. A typical use of this type of chart is to show the proposed and actual progression of the electrical installation. We can see immediately the required and actual condition of the contract at any time. This allows the electrician in charge, the main contractor and the client to be advised of the progress of the contract.

The client will want to know if the job is going to plan and if the work will be finished on time. They will also be interested in how the job is progressing. There will be other work and arrangements that will need to be made on their behalf and timing is important. Any problems that can be foreseen and the repercussions for the rest of the work and contract completion, can be easily identified. This allows everyone concerned to discuss and resolve, to the best possible advantage, a suitable solution.

The electrician in charge will need to be aware of any problems that may arise as soon as possible. For example if the number of men on the site is reduced due to sickness or holidays then the work could fall behind schedule. Regularly updating the bar chart record will show the actual progress falling behind the intended progress and allow prompt action to be taken before the matter becomes serious.

Alternatively the job may be ahead of schedule and so delivery of materials will be required before the date originally stated to prevent delays occurring. Of course the weather also plays a part in many contracting operations and so the work may be behind schedule through nobody's fault. In these conditions it may be necessary to remove some labour from the site for a while or to delay delivery of materials to the site. Contingency plans will then have to be made to bring the work back to schedule once conditions allow.

All of these and other site organisational problems may be easily identified if you produce and regularly update a bar chart for all the relevant activities on the site.

The main contractor will also need to know the progress of all the trades on site as they will be co-ordinating all the contractors and operations in all areas of the site. Any change in the rate of progress, of any job, will have a knock-on effect on all the jobs that follow. The site agent will need up to date information from all contractors, then should any delay occur that affects your work he will notify you. This information should then be added to the bar chart and any adjustment to the work rate of your team will be made as soon as possible.

The type of bar chart that we use will depend on the activity that we wish to observe. Figure 3.6 shows another variation on the theme and how bar charts may be used for easy monitoring of activities.

| Stages | Days | | | | | | | | |
|---|---|---|---|---|---|---|---|---|---|
| | 1 | 2 | 3 | 4 | 5 | 6 | 7 | 8 | 9 |
| First fixing conduit (lighting | | | | | | | | | |
| First fixing conduit (power) | | | | | | | | | |
| Complete conduit (lighting) | | | | | | | | | |
| Complete conduit (power) | | | | | | | | | |
| Wire conduit (lighting) | | | | | | | | | |
| Wire conduit (power) | | | | | | | | | |
| Fix luminaires | | | | | | | | | |
| Fit accessories | | | | | | | | | |
| Test | | | | | | | | | |

Work planned           Work completed

*Figure 3.6*

## Requisitions

It will often be necessary to produce a requisition for materials required from the site drawings that you are given. This process is used particularly on large sites where a record of materials issued from the central store, needs to be monitored and controlled. The same principles apply to the collection of materials for a small job. The main variables are the extent of the work and the reference material used, drawing, job sheet and the like. We shall examine one method of doing this here but your company may use one of their own. It is important to make sure that, whatever method is employed, a logical approach is used when preparing the requisition for materials.

The important thing is to make sure that we do not miss or forget any items. To do this some logical method must be used to enable us to list, check and request all the material required. Remember that we will not necessarily be able to work uninterrupted on site and inevitably someone will always need something from us whilst we are preparing our requisition. It is therefore a good idea to develop a process which deals with the requirements in stages and where the progress made can be identified easily.

If we consider a general case such as a conduit installation then we will require:

    First fix materials for conduit

    First fix materials for wiring

    Any second fix materials for conduit (box lids etc.)

    Second fix materials for wiring

    Special accessories

So if we divide our requisition into these main stages then we can deal with each stage separately.

We may also subdivide each stage to give relatively small sections – for example:

First fix conduit:

    Lighting
    (i) ground floor runs 1,2,3 etc.

    Lighting
    (ii) first floor runs 1,2,3 etc.

    Power
    (i) ground floor runs 1,2,3 etc.

and so on.

This provides us with a method of requisitioning materials that is in small stages so that work can be carried out efficiently and quickly and interruptions will not mean starting again from scratch.

There are special "take off" counters which operate rather like number stamps. Each time the device is depressed it increases the count by one and at the same time places a coloured mark on the drawing. This means that we can make a mark on each item, for example each twin socket outlet in an area, and keep a record of the number of items counted quickly, easily and reliably. These devices are particularly useful on large projects during the tender stage to establish quantity and hence cost. If such a device is not available then the use of coloured marker pens and a count notation system can produce similar results. A typical example of such a system would be coloured spot against each item with a coloured cross or number against every ten items, working up and down rows or working clockwise around the drawing. A good knowledge of the BSEN 60617 symbols will be a definite advantage here.

*Figure 3.7*

### Remember
Treat each take off or requisition as if the work is on the twentieth floor and there is no lift. Anything you have forgotten is going to involve a considerable amount of effort and inconvenience. We do not want to miss any item necessary to complete the work.

# Selection of tools and equipment

On the receipt of a specification for a contract the electrician in charge will have to assess the tool requirements for the work to be carried out. This will not include the day to day electrician's tool kit but any special tools that will be needed to carry out any specialist activity. We must therefore be able to recognise and identify the tool requirements for the work to be done.

In some instances this will be relatively easy. For example, if steel conduit is involved then a quantity of conduit benders, stocks and dies and pipe vices will be needed. In some instances however the requirements could be a little more obscure. The installation of a large sub main cable, which is to be terminated into a switch panel or bus bar chamber, may require the use of a hydraulic crimping tool to fit the terminal lugs to the cable ends. Any amount of MIMS cable work will require tools for terminations etc.

It is not only the type of tools and equipment that will be needed to do the work itself. Access equipment such as towers and ladders and any plant or power equipment, chasing tools, 110 V transformers and so on also need to be considered.

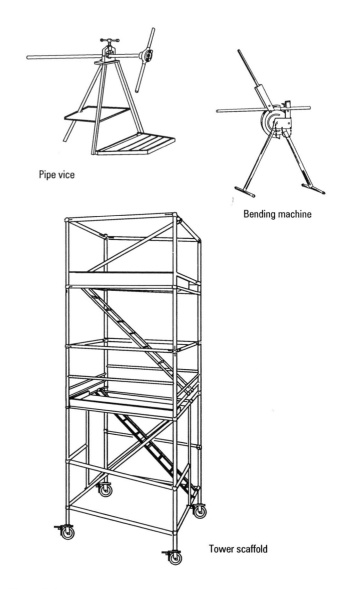

Pipe vice

Bending machine

Tower scaffold

*Figure 3.8*

The information contained in the specification will need to be used to assess these requirements. If the main contractor is responsible for all the chasing and cutting out then no equipment for this is needed. Should we need, for example, to fit conduit and fittings in a warehouse type of building then access equipment will be needed and a mobile tower may be the most effective.

By now we can see that our job involves a considerable amount of thought and consideration of aspects that many electricians take for granted. If we are responsible for the running of the site for the electrical work then we take the responsibility for ensuring all the needs are catered for.

## Try this

Using the drawing in Figure 3.9 make a requisition for the second fixing materials to complete the work.

Using the data supplied with Figure 3.9 compile a list of any special tools and access equipment required for the work involved.

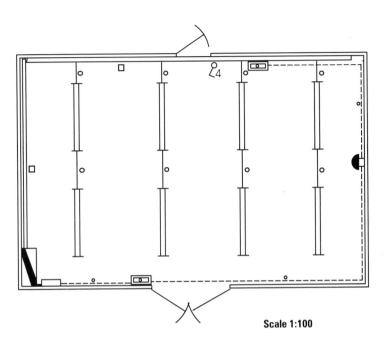

**Scale 1:100**

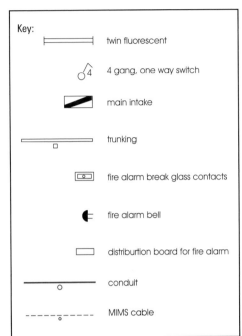

Key:

| | |
|---|---|
| twin fluorescent |
| 4 gang, one way switch |
| main intake |
| trunking |
| fire alarm break glass contacts |
| fire alarm bell |
| distriburtion board for fire alarm |
| conduit |
| MIMS cable |

*Figure 3.9*

**Data:**
Ceiling height – 3 m
Fluorescent lighting – steel trunking and conduit, PVC single cable
Fire alarm – MIMS cable

# Part 3

Having considered the documentation concerned with controlling and recording the activities on site, we need to look at the documents associated with the design and records for the electrical installation. We shall be looking at the requirements for a larger installation with a specification and drawings, but the same principles apply to the smaller work. Wherever we are responsible for the installation design we need to ensure the requirements of the client and any applicable standards are achieved.

During the production of the design for an installation the designer should produce drawings, schedules, charts and plans as part of the design process. These would be provided to the client as the "design proposal" and once these are agreed they are developed into a specification for the installation.

The responsibility for the design may form part of the contract awarded to the electrical contractor, in which case we may need to provide the design detail documentation, for the client to agree, before we commence work. In such cases the electrical contractor is responsible for the whole of the design.

The electrical contractor is responsible for the design of work where no specification is provided, such as smaller works, alterations and modifications and in particular work carried out for domestic consumers.

Where a specification is provided this may contain a varied amount of information, from a basic requirement in terms of layout and position of accessories and equipment, to a detailed design including cable sizes and routes. Dependant upon the extent of the information provided, and the terms of our contract, we may have to provide information for the design elements.

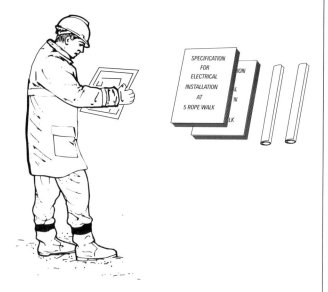

*Figure 3.10*

Once the design has been completed the information provided is used to construct the installation. Any drawings, specification, charts, plans or schedules which are to be used will be formally agreed and "issued" with a reference and date. These documents should be recorded and a register kept showing the details of the reference, date of issue, date received and status. A copy of the drawings will be used to construct the electrical installation and there are often changes to the requirements as the work progresses. This can result in revised information being issued.

When this occurs the register should be updated to show the receipt of new information, the existing information should be marked as "superseded", and the site operatives should be issued with a copy of the revised information.

We have a responsibility to ensure that the intended locations for equipment and fixed wiring are sensible, visually acceptable and co-ordinated with other services, equipment and the building structure. Any clashes should be referred to the design team, with any recommendations resulting from the site survey, in order to resolve the situation. Any recommendations as to how this can be achieved should be provided to the design team.

Following the completion of the electrical installation we have an obligation to provide the client with record drawings and information relative to the construction. These drawings and information need to be prepared during the construction process. Once the building finishes are applied it is usually impossible to accurately record the position of such items as cable runs which are concealed within the fabric of the building. The requirements for record drawings will be considered later in this section.

In order for the installation to be constructed as required by the designer and the client, we must follow the instruction given in the drawings, diagrams and specification. To do this we must be able to read and interpret the drawings and so we will look at the types of drawings and diagrams commonly used.

# Installation drawings

Our everyday work requires us to use information presented in the form of drawings and diagrams. This may be as a plan or layout to identify the location of components or fittings and accessories, or as diagrams to show the way in which a circuit is connected or how it functions. In addition to these we will also find diagrams used to give the sequence of equipment and controls.

To enable us to refer to these drawings with ease, a system of symbols that can be readily understood and easily interpreted is used. We must first establish what these symbols are and for which type of drawings and diagrams they are used.

## Use of BSEN 60617 symbols

The British Standards Institute issue BSEN 60617, a harmonized European Standard which contains standardised symbols. The use of these standard symbols makes referencing between different drawings much easier. In the contracting industry, for example, we will often be using drawings produced from several sources. If each of these uses their own set of symbols we will be constantly referring to the key to identify each item which may be time consuming and irritating.

If all drawings use the same symbols then it means that:
- we will become familiar with the everyday symbols and no referencing will be needed for the majority of work
- cross referencing will be made easier as each drawing will be using the same symbol for each item
- a key will only need to be drawn out for special symbols for unusual items, instead of every item used
- information will be taken from the drawing quicker and with a lot less stress

So the use of a standard set of symbols has a number of advantages. We shall consider the types of symbols that we are most likely to encounter.

## Location drawing symbols

These are used to indicate the type of outlet, fitting or accessories used. These symbols are only used on plan type drawings to show where in the building these items are to be located. Figure 3.11 shows some of the BS EN 60617 symbols used for this purpose.

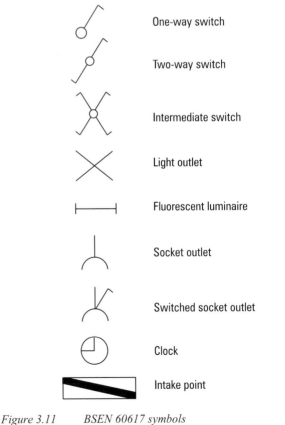

Figure 3.11    BSEN 60617 symbols

Of course this is not an exhaustive list and there are many others in common use. The full set of symbols can usually be found in the public library reference section. If we inspect these symbols more closely then we can see that some of them are made up of two or more symbols combined.

For example a distribution board has a symbol of:

Figure 3.12

A heating appliance has a symbol of:

Figure 3.13

So if we wish to show a distribution board controlling a heating load we combine the two symbols to form one as shown:

Figure 3.14

There are many cases where general symbols can be put together to make specific composite symbols.

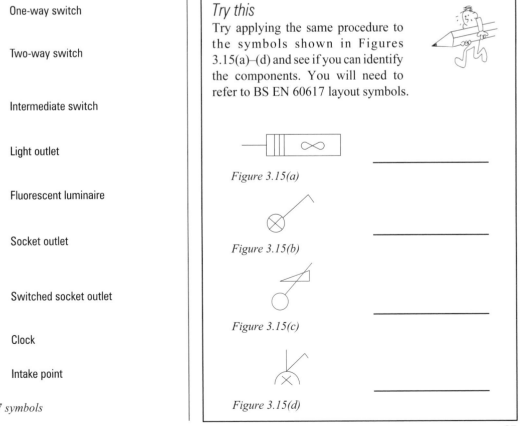

### Try this
Try applying the same procedure to the symbols shown in Figures 3.15(a)–(d) and see if you can identify the components. You will need to refer to BS EN 60617 layout symbols.

Figure 3.15(a)

Figure 3.15(b)

Figure 3.15(c)

Figure 3.15(d)

## Circuit diagram symbols

These are used to indicate the location of components within a circuit. Each component has a symbol to identify its function and they are used in a wide variety of circuits. Unlike our location symbols these are not normally shown on a scaled drawing and in many cases the components are not shown in their relative positions. Their purpose is simply to indicate the component and its function within the circuit. Figure 3.16 shows some of the BS EN 60617 circuit diagram symbols that are commonly used.

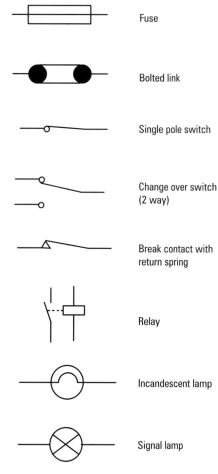

| | |
|---|---|
| | Fuse |
| | Bolted link |
| | Single pole switch |
| | Change over switch (2 way) |
| | Break contact with return spring |
| | Relay |
| | Incandescent lamp |
| | Signal lamp |

Figure 3.16

As before we will find that some of these symbols are made up of a number of others. As an example we can see that a symbol for a single pole switch is:

Figure 3.17

and that this can be used to produce symbols for a double pole switch and a triple pole switch (Figure 3.18) and so on.

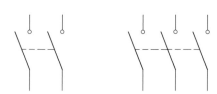

Figure 3.18        Double and triple pole switches

### Try this

Try applying the same ideas to the symbols shown in Figures 3.19(a), (b) and (c) and see if you can identify the component. You will need to refer to BS EN 60617 circuit diagram symbols.

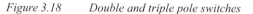

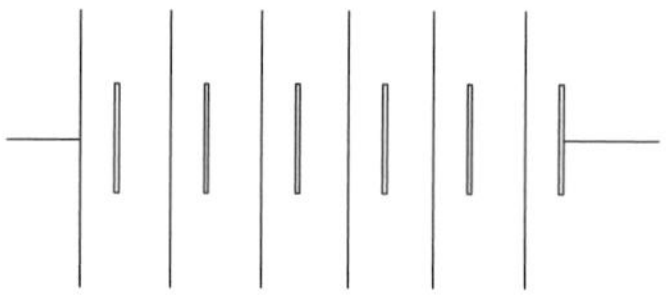

Figure 3.19(a)

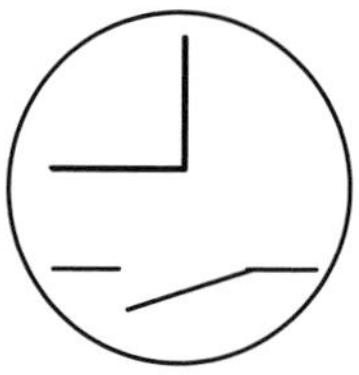

Figure 3.19(b)

Figure 3.19(c)

Now that we know the different types of symbols that we are most likely to see, i.e. the Location and Circuit Diagram symbols, we can consider the types of drawings and diagrams in which they are used.

## Layout drawings

These are used to indicate the particular location of outlets, accessories, components and so on. They are rather like an atlas indicating the relative locations of towns relative to one another. Drawings featuring these symbols will be to scale although the symbols themselves will not be scaled down. At this stage we are considering installation drawings so these will feature symbols from the architectural layout symbols from BS EN 60617.

It is normally the design engineer or architect who produces and issues these drawings to indicate the location of sockets, luminaires, switches and so on. The electrician on site will use layout drawings to establish the location of the accessories within the building and fit them accordingly. As the symbols are not to scale it is a common practice to take measurements to the centre of the symbol.

*Try this*

Figure 3.20 shows a simple layout drawing.

Construct a key to identify each type of component. You will need to refer to BS EN 60617 layout symbols.

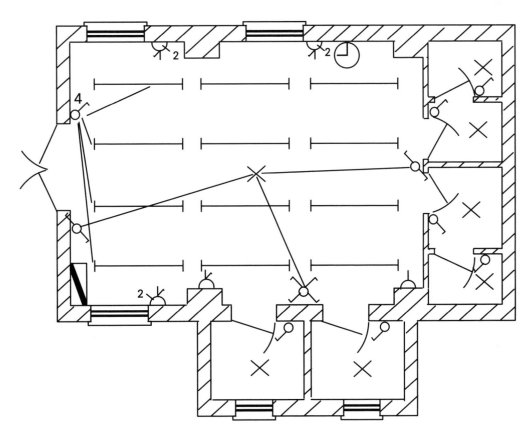

*Figure 3.20*

Key:

# Part 4

## Circuit diagrams

These are used to show how the components of a circuit are connected together, often in the simplest way and bearing no relationship to the actual positions. We can consider these to be the basic requirements for the circuit to operate correctly. A typical example of a circuit diagram is shown in Figure 3.21.

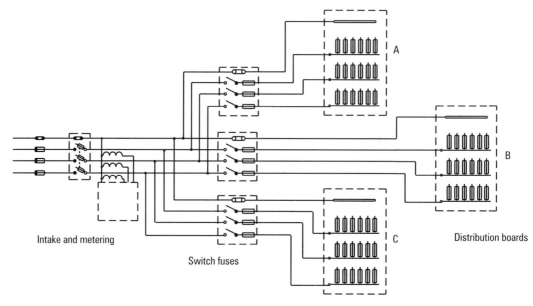

*Figure 3.21      Circuit diagram*

## Wiring diagrams

Wiring diagrams are used to indicate the locations of the components relative to each other and with regard to their actual geographical location. They will indicate the way in which the conductors are connected to each accessory and the actual number of conductors used. Their proposed route will also be indicated. An example of the wiring for our original circuit diagram is shown in Figure 3.22.

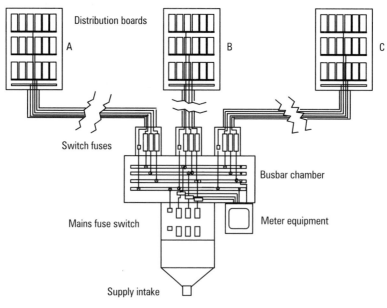

*Figure 3.22      Wiring diagram*

# Block diagrams

We use these to indicate the sequence of components or equipment. As the name suggests, they are just labelled blocks used to represent each item. They are linked by lines to show the sequence required. They are similar in many ways to a flow diagram used to indicate a sequence of events. Figure 3.23 shows a typical block diagram for an industrial main supply intake position.

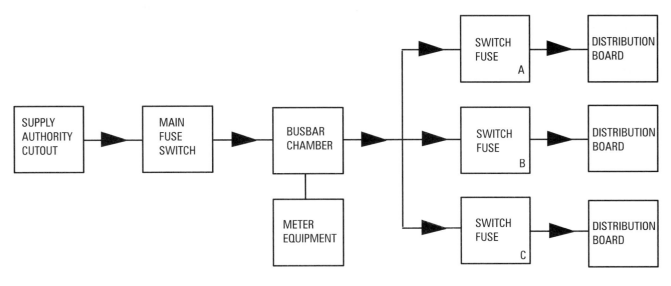

*Figure 3.23        Block diagrams*

# Record drawings

Often referred to as "as fitted" drawings these are produced from drawings marked up on site, usually on copies of the original site layout drawings, or using overlays. The purpose of these drawings is to indicate the exact routes taken by the wiring installed. They will show the location of the cables within the building by use of a single line and usually indicate next to this line the number of conductors or cables included at each location.

We use as fitted drawings to enable easy identification for repair, additions or access to cables for modification to wiring, and so on. Figure 3.24 shows a typical, simple, as fitted drawing.

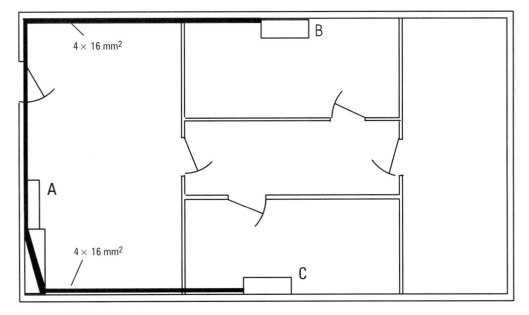

*Figure 3.24        Record drawing*

## Overlays

This is a term applied to a transparent sheet of tracing type paper that we lay over the original drawing. If we indicate a few location points so that the drawing and overlay can be aligned we do not need to reproduce the whole layout. We can then enter information on to the overlay for "as fitted" drawings and so on. This means that we need only one master drawing and we can create any number of overlays we like.

This means that we can separate the "as fitted" drawings for lighting, power, machines and the like, making the overlay easy to read. If an overall view of the entire system is required a number of overlays may be placed one on top of the other. This approach is particularly useful when comparing with other trades to check for problems with location of equipment or routes.

*Try this*

Using the layout drawing in Figure 3.25 prepare the following:

(a) A block diagram for the mains intake control gear showing the meter, main isolator and distribution.

(b) A circuit diagram for the complete lighting circuit.

(c) A wiring diagram for the two-way lighting circuit.

Using tracing paper prepare an overlay to give an example as to what the as fitted wiring could look like.

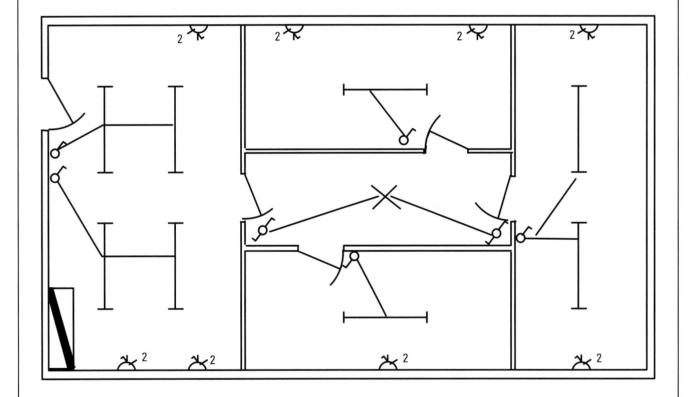

*Figure 3.25*

*Answers to Try this p.42:*
(a)

(b)

(c)

Documentation used to record on site activities plays a vital part in the control and efficient operation of any company. Such documents include time sheets, day work sheets and job sheets.

The recording of materials and equipment delivered to site and issued to the operatives is an essential part of the site control. The documents used to perform this function should be checked, signed and stored in a safe place for further reference.

It is important to programme and monitor the progress of work on site. Any changes to the rate of progress must be identified as soon as possible to minimise disruption and additional cost.

The issue, receipt and production of drawings related to the electrical installation are major considerations throughout the project. Those related to the construction of the installation need to be recorded and issued to the operatives on site, ensuring that the latest information is always used.

The drawings used to record the actual installation need to be produced as the work progresses and not left to the end of the job when much of the installation will be obscured from view.

## Self-assessment short answer questions

1.  List the documents required to monitor and control the following functions:
    (a) Work carried out by operatives on site
    (b) Materials and equipment delivered to site
    (c) Progress of work on site
    (d) Detailed information relating to the construction of the installation
    (e) Information on the completed installation

2.  List the minimum detail which should be recorded on a job sheet.

3.  Using the installation detailed in the "Try this" exercise (Figure 3.9), produce a bar chart for the first fix installation.

4. Using the installation detailed in the "Try this" exercise (Figure 3.9), produce a requisition for the first fix installation.

5. List the diagrams and drawings which would be required to carry out an electrical installation in a factory extension, from design to completion.

# 4

# Survey Requirements

Before you start work on this chapter, complete the exercise below to ensure that you remember what you learned earlier.

The completion of documentation forms a vital part of the _____ and _____ of progress on site. These documents include:

* time sheets and _____ _____ for the recording of the operatives _____.

* _____ _____ and invoices for the control and record of equipment and _____.

* layout drawings, _____ diagrams, _____ diagrams and _____ diagrams for the details of the installation

* _____ to provide details of the installation as it is installed

* programmes and ____ _____ to plan and monitor the activities on site

_____ and monitoring the activities on site are _____ to ensure that _____ materials, plant and _____ are available on site to maintain _____. Failure to do so may have serious cost implications for the project.

## On completion of this chapter you should be able to:

◆ identify survey methods and equipment and state the factors which affect their selection with reference to:
  – method
  – equipment
  – factors
◆ state the responsibilities of the installer, the site owner and affected parties in respect of undertaking a survey
◆ state the need for cooperation between the installer, the site owner/occupier and affected parties
◆ identify the assurance requirements necessary prior to survey with respect to authorisation and appropriate insurance

# Part 1

During the course of the preparation for the design and compilation of the record data for the completed installation we may need to undertake some site surveys. These will involve measurement, marking out and recording details. It may also involve access to areas of the site which may require security clearance, access arrangements and special access equipment. We shall consider the requirements for carrying out such surveys including the equipment and precautions which may be required. Before we can commence any survey there are a number of important areas which need to be addressed:

Access

Insurance

Authorisation

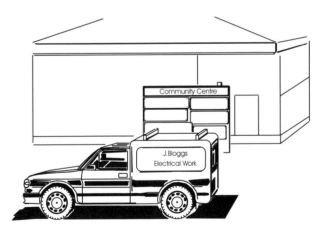

*Figure 4.1*

# Access

Before we can undertake a survey, we must obtain the cooperation of the site owner and the occupier in order that we can obtain access to the site. We need to confirm the details of the areas where we require access, and provide some indication of the extent of the work we are intending to undertake during the survey. The user of the site will need to know the duration and the likely disruption that will be caused in order to agree a suitable time for the survey to take place.

To provide this information in a meaningful way, dependant upon the extent of the work involved, we may need to provide a method statement for the survey activity detailing the relevant information.

We must also ensure that any third party, involved in activities which may be affected by our survey, is made aware of the proposed dates and duration where appropriate. Once the arrangements have been finalised all those parties affected by the survey should be advised in writing. This may be of particular significance if other work is being arranged which may be affected by our actions.

For example there may be a requirement to check the operation of an existing ventilation system which needs to be undertaken out of normal working hours. If this is arranged for the same time as our survey, and we need to isolate supplies for testing, then serious disruption could result. Whereas if we do not need to isolate supplies, perhaps we are only considering cable routes and co-ordination details, then it is possible that both activities could be carried out at the same time.

Similarly wherever we need to isolate supplies or equipment, we may interfere with the work of others. Likewise certain activities, such as major cleaning, painting, construction and demolition, could seriously disrupt our own work.

Figure 4.2    Getting in the way!

In addition to the need to arrange and agree the physical access to the site we must also ensure that the appropriate assurances are in place. This includes such requirements as:

- insurance
- authorisation
- agreed extent

Figure 4.3    Agreeing a survey method statement.

# Insurance

We need to confirm that appropriate insurance is in place to cover the activities we are to undertake and any events that may occur as a result of our actions. Contractors should have a public liability insurance to cover their everyday activities. These policies generally have some restrictions on the extent of cover and liability, and this may mean certain events are not covered. For example, loss of earnings or data to a third party as a result of interference with computer equipment may be an exclusion on the policy.

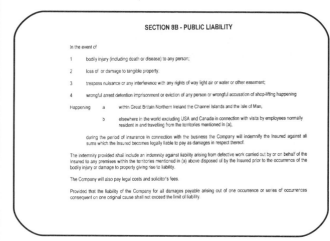

Figure.4.4    Part of a typical public liability policy

If we are to undertake a survey and tests in a building which has a heavy use of IT equipment we need to ensure that the liability is covered by our insurance. Inadvertently causing computer equipment to go "off line" may result in a claim for considerable compensation for loss of data, earnings, production and custom. Similarly we could find that other areas of activity are excluded from the policy. It is vital to ensure that the level of cover is adequate for both the work we are to undertake and any likely events which may result from that work.

## Authorisation

In some instances we may need to obtain authorisation for the work we are to undertake. If we are to have access to "sensitive" areas with particular security requirements or risks, we are going to need suitable written authorisation to carry out a survey. This may be due to the need to safeguard sensitive information or valuable stock, or as the result of a technical requirement, such as access to HV switch rooms and the like. In the case of the latter we would need not only the authorisation but also a permit to work.

It is essential therefore that we take all the necessary steps to ensure that the survey work can be undertaken at the appropriate time and date, with the minimum of inconvenience, and that the activities are suitably insured.

---

*Remember*
The installer, site owner, site user and any other affected parties have a responsibility to provide reasonable access and cooperation to enable the survey work to be undertaken. Each
should ensure, as far as they are able, that every effort is made to enable the work to be completed with as little disruption to the other as possible and to afford reasonable assistance to allow this to take place.

---

Having considered the necessary arrangements for the survey to be undertaken, we need to examine the detail of the survey itself.

A survey may fall into a number of categories dependant on the information which needs to be obtained. The greater the extent of the information required the more involved both the survey, and the methods required to achieve the end result, will be.

If we consider a survey of people using an electrical wholesaler during the course of a week, then the extent of the survey could be quite varied. It may be that a simple head count is required to establish how many people pass through the doors over the week. Alternatively it could be quite involved, collating details on the nature and size of the company involved, the quantity and content of the order, the time taken to deal with the customer, the total value of the order and all manner of other details.

So the type and extent of a survey are dependant upon what it is intended to achieve and the information that is to be collected. We shall consider some of the types of survey which are most commonly related to electrical installation work. Beginning with the;

## Visual survey

This survey, whilst beginning as a visual survey, often evolves to require some measurement or a least a counting exercise. A common example would be a survey to establish whether there are any adverse conditions involved along a proposed route for a new distribution cable (sub main) in an existing building. If no such conditions exist then there is probably no requirement other than the visual survey of the route. However if some adverse conditions are found, such as a high ambient temperature in a particular area, then measurement of the length of run and possible alternative routes would be necessary.

## Site access survey

The most common and simple surveys we are likely to be involved in is to establish physical access to the site, or a suitable location for the set up of the storage and office facilities. We carry out such surveys, in their most basic form for every job we undertake. For example, considering items such as where to park the van, which access door to use, where to put our tools and so on. For larger projects this generally involves ensuring that adequate site access is available to allow equipment and material to be delivered to the site. Once delivered can these be moved to the required area for storage and use, and is any special equipment required for this activity? This is normally a visual survey, with some measurement required to determine the size of access doors, storage areas and clearance distances along the routes to be used.

*Figure 4.5        Access measurement being undertaken.*

The purpose of this exercise is to ensure that it is possible to undertake the intended work. During the course of the survey any areas of difficulty should be identified and where possible alternative solutions should be considered. In some instances it may be necessary to carry out building modifications or provide additional equipment before the work can commence. For example it may be necessary to have some internal doorways enlarged temporarily to allow access for equipment. These would be reinstated on the completion of the work. Alternatively access through an occupied building to get material and equipment to an upper floor may prove so impractical that an external hoist needs to be installed and access constructed using a window opening (Figure 4.6).

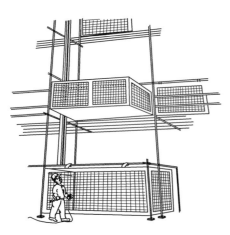

*Figure 4.6        External hoist arrangement.*

A further example of such a survey would be where, in order to carry out some additional work, items of equipment have to be removed and stored until they are refixed. In this case a count of the items of equipment would be necessary and the size of each established. This then enables a calculation to be carried out to determine the time required to dismantle, the space required to store the equipment and the need for any additional items such as packaging, protection and transport.

## Location, costing or manufacturing survey

Where detailed drawings are not available a more thorough survey may be required in order to establish requirements for quantities of materials for installation. Specialist installers may need to carry out detailed surveys with accurate measurements in order that bespoke equipment can be produced to meet the space confinement of the location, such as main control panels and the like. It is common for a detailed survey to be carried out to establish quantities of materials actually installed. These are usually required in connection with work where payment is made against material costs, such as additions to a contract or work paid on a "time and material" basis. Such surveys would require a schedule of quantities to be produced, in much the same way as the take off for tender or ordering which we considered in the previous chapter.

## Structural survey

Wherever we are carrying out an installation in an existing building, some degree of structural survey will be required. The extent of the survey will depend on the nature of the building and the decorative finish. Without prior knowledge it may not be possible to tell from a visual inspection whether an internal wall is constructed of brick, blocks or timber or whether it is plastered or dry lined. We are usually obliged to carry out some survey, however small, to establish the nature of the building fabric before we commence work.

In connection with the electrical installation it may be necessary to carry out a structural survey to establish whether areas are suitable for carrying the required loads. Lightweight fibre or plasterboard partition walls may not be suitable for mounting distribution boards, cable tray runs and heavy equipment. If the structure of the building does not lend itself to the proposed installation method then alternatives will need to be considered. This could involve a change to the type of installation or the use of alternative support methods. For example, we may need to install supports down from a concrete ceiling as opposed to wall mounting equipment.

These structural considerations and the associated fixings will be considered later in this book.

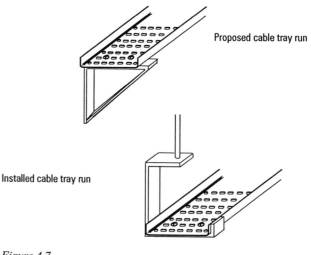

Proposed cable tray run

Installed cable tray run

*Figure 4.7*

It may also be necessary to carry out a structural survey in order to determine whether an existing building structure is adequate for the proposed additional work. We may be asked to install plant or equipment in a particular location within a building and it is necessary to ensure that the structure will support the weight of the equipment. This may be particularly significant with equipment such as generators, UPS equipment, chillers or air conditioning equipment and water storage vessels. It is usual in such cases for a structural engineer to be engaged to carry out the major structural surveys.

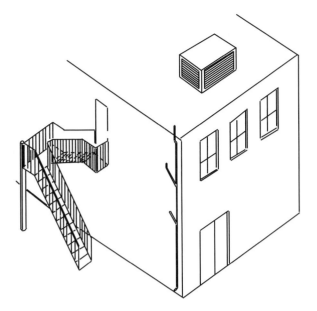

Figure. 4.8    *Large generator mounted on the roof of a building.*

The extent of access or structural modifications or the need for particular installation techniques to be used can have a considerable effect on the cost of the work. Enabling and reinstatement work is essential but provides no added benefit for the client. Similarly a more costly installation technique, as a result of the building structure, results in no advantage or visible change for the client. These requirements need to highlighted early in the design stage of the project so that a true reflection of the work involved can be presented.

Now we have considered the type and extent of the survey which may could be required we shall examine the equipment which may be involved in carrying out the survey.

# Part 2

# Equipment

As we have already stated most surveys involve some amount of measuring so inevitably we need to consider measuring devices. In order to adequately carry out the survey we shall also need to access the areas to be surveyed and we therefore need to consider access equipment. We will begin by considering the measuring equipment.

## Measuring equipment

Basic measuring equipment such as rules and tapes were covered in books in the foundation course of this series and as such will not be covered again in any depth here. However there are some additional items of measuring equipment which are worthy of consideration at this time.

## Digital tape

Developments in micro technology have enabled the production of a steel tape with a built-in digital readout of the measured length. Many of these can provide details in metric or imperial measurement, automatically add the width of the case to the measurement, provide memory recall and add successive measurements to arrive at a total. These devices are particularly useful in locations where it is difficult to read the tape, either due to poor light or positioning of the tape and for the addition of successive measurements.

Figure 4.9    *Digital tape*

## Range finder or digital estimator

These devices employ a laser to take measurements and can calculate area and volume providing a digital readout of the figures. These devices are usually very accurate, within a range of say ±0.5% to ±0.01% for the linear measurement. These are useful in large, clear areas where accurate length, area or volume measurements are required as the operation can be carried out from a single location without the need for long tapes and steps. Complete calculations can also be carried out providing the required data in a single exercise. They are particularly useful for calculating the volume of, say, a proposed storage area on a site being surveyed.

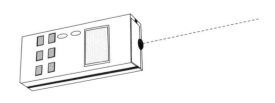

Figure 4.10    *Range finder*

# Lasers

The development of laser measurement and levelling equipment has made a large range of very accurate devices to aid both survey and construction available at an affordable price. Although there are many produced, we shall consider only two types here as most are variations on a common theme.

## Plumbline laser

These devices are designed to provide accurate vertical references and typically cover a range of about 40 metres and incorporate a self levelling mechanism. This mechanism usually involves a pendulum device within the laser to ensure that, irrespective of the surface level, the laser automatically levels providing accurate vertical references and are generally accurate to around 0.01%. Most offer both a line or dot configuration allowing easy transfer of measurement from floor to ceiling when surveying or marking out. They also offer accurate vertical references over fairly long distances.

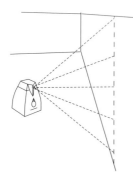

*Figure 4.11*     *Plumbline laser*

## Laser level

This comprises a spirit level with a built-in laser which can be mounted on a tripod with the facility to level and adjust the height of the device. They are generally provided with a "target plate" for fixing the exact location of the laser height. The use of such a device allows the setting out of equipment within an area to the same height, irrespective of the terrain level, and avoids the human error inherent in the transfer of measurement. A similar result can be achieved using a water level. However, the laser level does help to eliminate any transfer errors. The laser level is typically accurate to around 0.05% providing a very precise level facility over a typical range of a 40 m radius.

*Figure 4.12*     *Laser level and target plate*

## Measuring wheel

Finally we need to consider the measuring wheel which is particularly useful for measuring long runs on level ground. The wheel is generally constructed with a circumference of one metre and is fixed to a handle and digital counter. One revolution of the wheel is therefore a linear distance of one metre and it is recorded on the digital counter, rather like the milometer fitted to a car. This enables quick and accurate measurement of length where the ground is fairly level and smooth. The application could be for lengths of run of lighting trunking, wall mounted trunking or lengths of cable. Car park or roadway lighting cable measurement is a common application of a measuring wheel.

*Figure 4.13*     *Measuring wheel*

*Try this*

List the device most appropriate for carrying out the following tasks

(a) Positioning the mounting heights of a row of outlet plates for computer socket outlets during the 1st fix stage of construction.

(b) Marking out the fixings for a vertical run of bus-bar trunking.

(c) Measuring the length of cable required to supply lighting columns in a car park.

(d) Measuring the area of a storage facility

Having reviewed some of the types of measuring equipment which may be appropriate, we now need to consider the access equipment which we may need. It is important to remember that "access" does not simply mean the means of reaching a working height. It is used to refer to access to "carry out the work". A scaffold may be used to provide access to carry out the installation of an illuminated sign across the entrance to shopping mall for example. The scaffold platform is the means of access to carry out the sign installation but a ladder or hoist will provide access to the working platform. In this book the reference to "access" encompasses all the elements of access to the working platform and the platform itself.

# Working at height

Falls at work account for some 50% of fatalities and most accidents involving falls could have been prevented if the right equipment had been provided and used. It is, therefore, important that the appropriate precautions are taken to prevent falls during the course of the survey and the subsequent installation and commissioning work. The basic requirements for simple access equipment, such as ladders and steps, have been covered in the foundation course series of these publications "Starting Work", and it is not proposed to cover these again here.

It is worthwhile to consider the basic precautions and actions to be taken before working at height. This information allows a system of checks to be considered and applied before undertaking such work, whatever the circumstances.

When selecting the means of access it is important to remember that;
- ladders should only be used as a means of access when it is safe to do so
- ladders should only be used as a means of access for short periods of time and prolonged working from ladders should be avoided
- where possible use a scaffold tower or mobile working platform
- when working on a platform, guardrails and toe boards should be fitted before work commences
- safety harnesses should be worn
- harnesses and arrester nets should only be relied on as a last resort when it is not possible to provide a suitable working platform, such as when hand rails are removed to allow the loading of material.

> ## Remember
>
> When a ladder is in position it must follow the 4:1 rule. This means that the height of the ladder above the ground must be four times the distance the ladder is out from the foot of the wall. The ladder will then be at an angle of 75° to the ground.

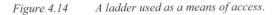

*Figure 4.14     A ladder used as a means of access.*

> ## Remember
>
> Harnesses and lines or safety nets do not prevent falls and they only provide protection for the individual using a harness in the event of a fall.

Before commencing work at heights it is important to check;
- there is a safe method of getting to the work area
- the equipment is suitable for the site and conditions
- equipment erected by others is safe for use before commencing work
- persons responsible for operating powered access equipment are suitably trained and qualified to do so
- hand rails, toe boards, harness rails and other safety provisions are in place

> ## Remember
> Each individual has a responsibility for their own safety and that of others who may be affected by their actions. Ensuring that a safe means of access is provided and the necessary safety equipment is supplied and used before you commence work is your responsibility.

It is important to remember that working at height does not simply involve work high above ground level. The use of a pair of steps or a simple hop-up constitutes working at height. In addition working at height also applies to work carried out adjacent to excavations. Suitable protection should be provided to ensure that people working at ground level do not accidentally fall into excavations, open manholes and the like.

Figure 4.15    *Providing a safe means of access also applies to access over trenches, manholes and the like.*

More detail on the requirements for working at height and the appropriate precautions that should be taken can be found in the Health and Safety Executive guidance HS(G)150, Health and Safety in Construction. It is advisable to obtain access to a copy of this publication and to become familiar with the requirements for the protection of yourself and others.

*Try this*

When carrying out a survey there is a need for co-operation from the parties involved. Identify the four main parties who would have some responsibility to ensure the survey may be carried out.

Points to remember ◀ – – – – – – – – – – – – –

During the course of the design and record stages of electrical installation work there may be a need to undertake surveys.

These fall into three main categories, visual, detailed and structural.

It is important to ensure that suitable access to the areas of the site involved are agreed and any special requirements are catered for.

The company carrying out the survey work needs to ensure that they have suitable insurance cover for the work to be undertaken and any likely incident as a result of the survey.

Before the survey work is undertaken the precise nature of the work involved must be detailed.

Checks need to be made to ensure that the work does not interfere with other programmed works and other work will not interfere with the survey work.

Consideration must be given to the means of access for plant, equipment and material during the construction stages of the work.

Suitable equipment should be used to ensure that the survey produces accurate information which is correctly recorded.

Suitable access equipment should be used at all times to allow the work to be undertaken safely.

1.  List the two main requirements prior to the commencement of a survey.

2.  List the equipment which may be required to undertake a survey for the installation of an additional machine to an existing factory installation.

3.  List the factors, relevant to the location, which may affect the survey or the required equipment.

4.  During the course of a survey there will be data to be collected and recorded, explain why the accurate recording of the data is important to the
    (a) design and
    (b) costing of the work to be carried out.

# Progress Check

**Circle the correct answers in the grid at the end of the multi-choice questions.**

1. The organisation which represents the interest of the employer in the electrical contracting industry is the
   (a) AEEU
   (b) JIB
   (c) ECA
   (d) NICEIC

2. The terms and conditions agreed between employer and employee form part of a
   (a) national agreement
   (b) grading scheme
   (c) national register
   (d) contract of employment

3. The success of any project depends upon
   (1) good communication and co-operation and
   (2) client's agreement to any proposals
   (a) statement 1 is correct and statement 2 is incorrect
   (b) both statements 1 and 2 are correct
   (c) statement 1 is incorrect and statement 2 is correct
   (d) both statements 1 and 2 are incorrect

4. One important aspect in the design process is
   (a) researching and collating information
   (b) ensuring the cost is kept to a minimum
   (c) checking delivery dates
   (d) establishing the site set up

5. A statutory regulation is one which
   (a) does not change
   (b) provides guidance
   (c) is enforced by law
   (d) is agreed by the contract team

6. The general regulations covering safety in all places of work are
   (a) Electricity Supply Regulations 1988
   (b) The Electricity At Work Regulations 1989
   (c) The Health and Safety At Work Act 1974
   (d) The Construction (Design and Management) Regulations 1994

7. (1) Health and safety at work is the responsibility of every employer and employee
   (2) Your responsibility for health and safety can be passed to your supervisor
   (a) both statements 1 and 2 are correct
   (b) statement 1 is correct and statement 2 is incorrect
   (c) both statements are incorrect
   (d) statement 1 is incorrect and statement 2 is correct

8. Which of the following **does not** normally warrant any special precautions under the COSHH Regulations 1994
   (a) adhesive for PVC conduit
   (b) cutting paste
   (c) barrier cream
   (d) fibre glass

9. Under the requirements of RIDDOR notification must be provided to the HSE regarding
   (1) serious or fatal accidents and
   (2) in the event of work related diseases.
   (a) both statements 1 and 2 are correct
   (b) statement 1 is correct and statement 2 is incorrect
   (c) both statements 1 and 2 are incorrect
   (d) statement 1 is incorrect and statement 2 is correct

10. The status of the Building Regulations are
    (a) guidance for developers
    (b) non-statutory regulation
    (c) statutory regulation
    (d) regulation for new construction work only

11. Regulatory and planning requirements may have implications for the electrical contractor. Which of the following lists the four main areas affected
    (a) finance, legislation, resources and technical requirements
    (b) finance, labour, resources and technical requirements
    (c) finance, legislation, transport and technical requirements
    (d) finance, legislation, resources and plant

12. Where work to be carried out has implications for the health and safety of operatives and those around them, the activity should be carried out under
    (a) a code of practice
    (b) a permit to work
    (c) a method statement
    (d) increased supervision

13. The detail recorded on an operative's time sheet will **NOT** normally include
    (a) address of the work
    (b) number of hours worked
    (c) the material used
    (d) the name of the employee

14. When only some of the total number of items ordered are required on site to meet the programme, the delivery note accompanying the goods should show the status as
(a) completion order
(b) outstanding order
(c) part order
(d) finalisation order

15. The take off for the materials required for a particular area of the work would be carried out using which type of diagram
(a) block
(b) circuit
(c) wiring
(d) layout

16. The diagram which details the position and function of the components within a circuit is known as
(a) wiring diagram
(b) layout diagram
(c) circuit diagram
(d) block diagram

17. Before beginning a site survey the contractor must obtain the consent and co-operation of
(1) the owner of the building and
(2) the occupier of the premises
(a) both statements 1 and 2 are correct
(b) statement 1 is correct and statement 2 is incorrect
(c) both statements 1 and 2 are incorrect
(d) statement 1 is incorrect and statement 2 is correct

18. In order to carry out a site survey contractors should ensure that they have the appropriate
(1) tools
(2) access
(3) plant
(4) equipment
(5) insurance
(a) only items 1, 2, 3, and 4 are required
(b) only items 1, 2, 3, and 5 are required
(c) All of the above are required
(d) only items 1, 2, 4, and 5 are required

19. 50% of the accidents resulting in fatalities involve
(a) falls
(b) electricity
(c) gas
(d) explosion

20. Working adjacent to trenches and earthworks is subject to the same requirements as working
(a) in a confined space
(b) at heights
(c) underground
(d) under a permit to work

*Answer grid*

| | | | | | | | | | |
|---|---|---|---|---|---|---|---|---|---|
| 1. | a | b | c | d | 11. | a | b | c | d |
| 2. | a | b | c | d | 12. | a | b | c | d |
| 3. | a | b | c | d | 13. | a | b | c | d |
| 4. | a | b | c | d | 14. | a | b | c | d |
| 5. | a | b | c | d | 15. | a | b | c | d |
| 6. | a | b | c | d | 16. | a | b | c | d |
| 7. | a | b | c | d | 17. | a | b | c | d |
| 8. | a | b | c | d | 18. | a | b | c | d |
| 9. | a | b | c | d | 19. | a | b | c | d |
| 10. | a | b | c | d | 20. | a | b | c | d |

# 5

# Site Structures

Before you start work on this chapter, complete the exercise below to ensure that you remember what you learned earlier.

Before commencing any site survey you need to ensure that you have access, _____ and _____ to carry out the work.

The method which is used to carry out the survey will depend upon the _____ of the survey to be carried out. This may be as simple as a _____ survey, a detailed survey involving _____ , or it may need a _____ survey to determine the suitable siting of _____ and heavy equipment.

Dependant upon the extent of the survey we may need to use a range of _____equipment.

You may use measuring _____ , rules or _____ , to carry out the measurement but electronic devices such as laser _____ and _____ or digital _____ _____ and estimators may prove particularly beneficial.

Precautions need to be taken when working at height to ensure a suitable _____ to the working site and a safe _____ _____ from which to carry out the survey.

## On completion of this chapter you should be able to:

◆ state the importance that site structures and fabrics, and their characteristics, have on electrical installations
◆ state the importance of the interaction between the site structures, and fabrics, and the electrical installations
◆ state the external factors that affect your choice of wiring system
◆ state briefly the effect of these conditions on your choice
◆ use these factors to determine suitable systems for given locations
◆ state the safety procedures to be adopted when undertaking a survey

# Part 1

# Site structures

In this chapter we shall be looking at the external factors that affect the type of wiring system we use. These factors relate to the environment, construction and use of the building in which the electrical installation is to be constructed. These exterior factors may have a considerable influence on the type of installation or materials we use.

The factors we need to consider include

• the type of building construction
• the environmental conditions that prevail
• the purpose for which the building will be used
• any special considerations

We must consider these factors in a little more detail so we will be looking at each in turn beginning with:

## Type of building construction

The way in which a building is constructed and the materials from which it is built can have a considerable influence on the type of wiring systems we choose to install.

The main areas of concern in regard to the building structure are as follows:

### Is the material combustible?
If so then we must consider this in our selection of the type of cable and enclosures to use. It will also have implications regarding the temperatures produced in the electrical system during operation and fault conditions.

### Are the size and shape of the building such that it will allow a rapid spread of fire?
Again this will affect the types of cables and enclosures used. It will also require consideration when locating equipment.

*Is the building of considerable length or built on unstable ground which could result in the movement of one part of the building with respect to the other?*

If this is the case, the system we use must be able to accommodate this movement without damage or deterioration.

*Does the building contain movable partitions or false ceilings or is the structure completely flexible such as a tent or marquee?*

These types of installation require special consideration. Structures that contain false ceilings and the like offer space for the installation of wiring systems which would otherwise have been impractical in a building with a concrete raft ceiling. Buildings with movable partitions require a wiring system with sufficient flexibility to readily allow alterations to suit changes in building use.

Totally flexible structures such as tents are often used for functions and require lighting and power supplies. It would obviously be totally impractical to carry out such an installation using steel conduit or trunking. In any event these factors will all have some bearing on the type of wiring system we install.

# Environmental conditions

Having given some consideration to the type of building construction, we must also consider the environmental conditions that are likely to exist and how these will affect our choice of wiring system. We shall consider each of these points with a brief explanation as to their implications.

*Remember*

Environmental factors may be internal or external to the building.

## Temperature

High or low temperatures will have an effect on most modern materials used in cable production. Many types of cable are produced so we select the one best suited to the conditions that may exist. The temperature will also have an effect on enclosures and equipment and therefore we must select accordingly. For instance, we may decide that PVC conduit is not suitable for installation in a cold store, due to the low ambient temperature.

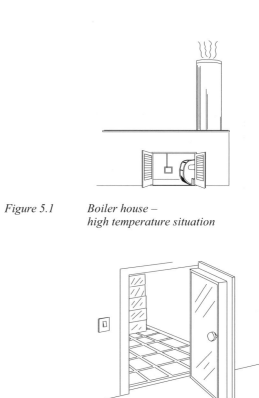

*Figure 5.1*    Boiler house –
                high temperature situation

*Figure 5.2*    Coldroom –
                low temperature situation

## Humidity

Having given consideration to the temperature that may exist we must also make allowance for the humidity, i.e. the water content of the air. As an example a greenhouse will probably have a high percentage of water in the air compared with, say, a furnace room. So whilst the temperature may be high in both cases the same type of installation may not be suitable due to the different levels of humidity.

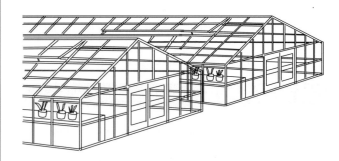

*Figure 5.3*    Electrical installations in greenhouses have to
                be suitable for the environment.

## Water

By this we mean is there likely to be water present in the area. In general this can be viewed in terms of water droplets and how they occur. In an area where washing down with a pressure hose takes place then the system will need to be designed to withstand water under pressure. In an area where a fine mist of water is likely to occur the conditions will be less onerous and so a different system may be employed. Special consideration needs to be given to equipment which is to be submerged in water, such as sump pumps.

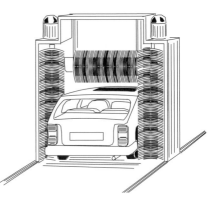

*Figure 5.5    Car washes and street lighting have to be wired in cables suitable for wet conditions.*

## Foreign bodies

We can look at this as the amount of dust and debris that will be present in the air. A computer room would often be considered a dust free environment whereas a builder's warehouse would not be and so different types of wiring system and accessories would be used for each.

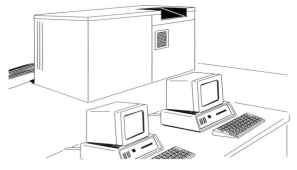

*Figure 5.6    Where computers are used dust is kept to a minimum.*

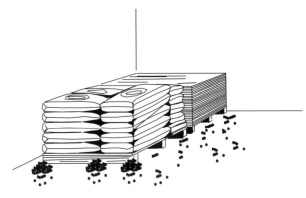

*Figure 5.7    Builders' stores tend to be dusty environments.*

## Use of the IP code

The IP Code for ingress protection is one method of checking to see if equipment is suitable for any particular area of risk.

A piece of equipment is coded by the use of two numbers and optional additional and supplementary letters if necessary.

## IP Code

The first part of the equipment reference is the code letters IP indicating that the following numbers and letters refer to those contained within that code.

The first characteristic number then identifies the degree of protection against solid bodies, covering a range between no protection at all and dust tight.

The second characteristic number indicates the degree of protection against the ingress of moisture, covering a range between no protection and continuous immersion in water.

The next characteristic is the "additional" letter related to the degree of protection to persons from access to "hazardous parts", covering a range between protection against contact by the back of the hand and protection against contact with a wire of diameter 1mm and length 100mm. It is only used where the protection against access to hazardous parts is higher than the initial number or where protection against access is provided but no protection is afforded to general ingress.

The final characteristic is the "supplementary" letter. This is provided from the product standard and refers to particular aspects of the equipment such as suitable for High Voltage and use under specified weather conditions.

Where a characteristic numeral is not required to be specific, it can be replaced by an X, or XX where neither number is required to be specific. If the additional and/or supplementary numbers are omitted then no substitution is required.

If we consider a length of bare copper overhead line, the IP Code would be IP00 (or IPXX) as no protection against solid objects or liquids is given. We can see from the tables in the IP Code that the higher the index number the greater the degree of protection offered.

The information given in the table is self-explanatory. Some equipment may be coded by use of the picture representation shown in the table. If in doubt refer to the IP Code BSEN 60529 (BS 5490).

## The IK Code

BS EN 50102 "Degrees of protection provided by enclosures for electrical equipment against external impact" contains information regarding the mechanical strength of enclosures referred to as the IK Code. This code identifies the general requirements and specific details will be contained with the product standard for the particular enclosure concerned.

The information is presented as the code letters IK, followed by a group number between 00 and 10 and these refer to the energy impact withstand value in Joules. For example IK07 indicates the enclosure has an energy withstand of two joules whilst IK00 indicates there is no specific protection specified.

IEE Guidance Note 1 contains details of the requirements of the IP code.

---

*Try this*

1. A motor is to be installed in an area which is prone to condensation and has people working in close proximity. State the minimum IP code number that would be suitable for the motor enclosure in these circumstances.

2. An enclosure has an IP of IP31. Give a brief description of the environmental conditions for which this enclosure is suited.

---

# Part 2

## Corrosive or polluting substances

In any area where such substances are likely to be present we must take particular account of the materials used and whether they are suitable for such an environment. Plating and cleansing activities often involve the use of such substances and thus will require special consideration including the use of anti-corrosive finishes, the use of special extract fans and filters.

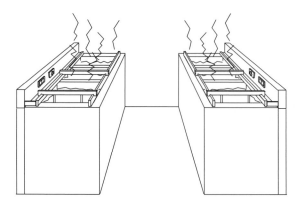

Figure 5.8    Electro plating baths can give off corrosive fumes.

Figure 5.9    If the correct finish for the environment is not used on steel conduit and trunking then they will corrode and no longer offer protection for the conductors. Should the enclosure form the cpc for the circuitry this corrosion may cause serious earth continuity problems.

## Hazardous areas

Special considerations should always be given to hazardous areas that could cause an explosion. Areas such those used for paint spraying, where there are high concentrations of flammable dust or fibre and also areas where petroleum is dispensed, must be wired using suitable materials and equipment. It is important to keep the explosive dust or vapours away from electrical equipment where a spark could cause ignition.

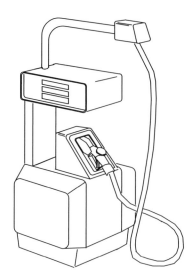

Figure 5.10    An example of a hazardous area.

## Flora or mould growth

As with the consideration given to corrosive substances many common plants and moulds produce corrosive chemicals. They may attract wildlife and often retain moisture so we must give consideration to these points. In some cases the application of chemicals to promote growth or kill pests may also be an important factor.

## Mechanical impact

In a lot of industrial premises we are, for example, installing cables in locations where they may be subjected to some form of mechanical impact. This situation may also occur in the most unlikely of places. The domestic installation with a cable run over a skirting board is a good example and we must consider the need for mechanical protection in such cases.

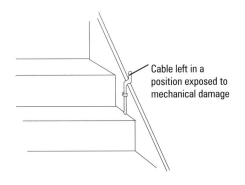

Cable left in a position exposed to mechanical damage

*Figure 5.11*

## Vibration

Where a wiring system is connected to a machine there is likely to be some vibration. As most fixed wiring systems are not intended to move we must make an allowance for this. A suitable flexible connection should be incorporated between the machine and the fixed wiring of the installation.This flexible connection should absorb the vibration and prevent damage to the fixed wiring, the associated terminations and connections. For example various forms of flexible conduit and MIMS cable with anti-vibration loops can be installed.

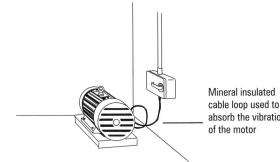

Mineral insulated cable loop used to absorb the vibration of the motor

*Figure 5.12*

## Solar radiation

If we intend carrying out an installation where some part is exposed to the effects of sunlight then we must ensure that the system of wiring and the materials used are suitable for such exposure. Tough rubber sheathed (TRS) cable, for example, is prone to rapid deterioration if exposed to direct sunlight as are some PVC compounds.

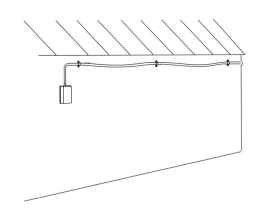

*Figure 5.13*    *PVC conduit exposed on an outside wall can be heated by the sun to become hot and sag between fixings.*

## Lightning

If part of the system is located outside of a building then we must consider the possibility of a lightning strike to the system and offer the necessary protection to prevent this from creating a hazard to both the system and the users. Installations supplied by overhead lines are a good example of such a case.

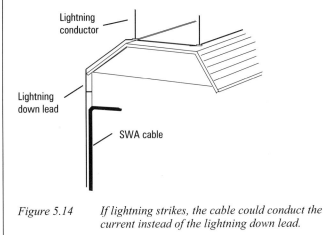

Lightning conductor

Lightning down lead

SWA cable

*Figure 5.14*    *If lightning strikes, the cable could conduct the current instead of the lightning down lead.*

## Electromagnetic effects

This covers a number of areas but they are all concerned with the effects of stray electric currents or electromagnetic radiation including electrostatic effects. The solution will obviously vary but may be as simple as installing screened cables, altering the routes of cables or relocation of equipment within the building.

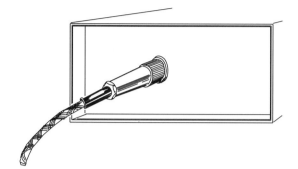

*Figure 5.15    Screened cables are used to shield the conductors from the effects of other electric fields*

## Temporary installations

There are no reductions in standards for temporary installations but different wiring systems are often used. Temporary lighting on construction sites, like other supplies, may be at 110 V to improve safety. A festoon lighting system is often adopted to give simple wiring and adequate all-over illumination. Factors such as fixing heights and distances apply here the same as on any other installation.

A 110 V power distribution system is used on larger sites and installed to give safe working voltages from a safely installed system. Some installations are not connected to the supply company's system but have their own generator. They may supply heavy power equipment such as tower cranes.

*Figure 5.16    Temporary installations can be supplied through transformers from the supply company's mains or from generators on site.*

In addition to these considerations we must also have some regard for the effects of natural phenomena such as wind and ground movement. As you can see this is quite a list of points to be considered but we've not quite finished yet. We must also take into account the buildings probable use.

## Purpose of the building

When we make an allowance for the use of the building, it does not simply cover issues like is it a paint store or a computer suite?

We must be aware of other factors such as:

**The capability of the people using the building**, for example are they able-bodied with full mobility or are they elderly people with limited mobility or speed of movement? It may be that the people using the finished building are disabled with requirements for wheelchairs and other mobility aids.

*Figure 5.17*

Another consideration is **the density of occupation**. Is the building one with a low density of people and therefore easy to evacuate in the event of an emergency or is it densely occupied with limited access thus making evacuation in the event of an emergency difficult?

We must also look at **the probability of people coming into contact with earth potential within the building** and from this determine the protection method(s) that may be required.

Well, that has just about covered all the main points that we need to consider that are external to the actual electrical installation. You can see that a degree of common sense will stand you in good stead when considering the external influences which will affect the electrical installation.

# Part 3

We must now give some thought to the structure of the building and how the types of structure and materials used will affect the electrical installation. Before we begin, it may be advisable to recap the terms used in connection with materials and what properties those terms indicate.

## Strength

The strength of a material is its ability to withstand an external force. Materials used in construction are often used, or specifically designed and constructed, to provide load-bearing strength. Steel or reinforced concrete beams are often used to construct floor supports, bridges and the like because of their load-bearing properties. The load-bearing property of a material is generally its ability to take a considerable weight without distortion or damage.

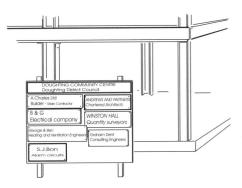

*Figure 5.18*

## Hardness

Hard materials are generally difficult to work with and the cutting and shaping of hard materials may require special tools. Some materials are worked on and then hardened to make them more hard wearing, others are naturally hard and may often be quite brittle. Hard materials and metals are typically used for the manufacture of cutting tools.

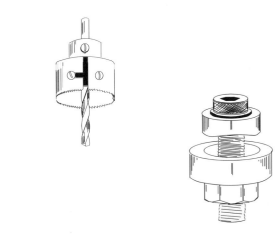

*Figure 5.19*     *Hole saw and hole punch manufactured from hardened steel*

**65**

## Toughness

The toughness of a material could be described as its ability to withstand repeated blows without breaking. This does not imply that the material will not change shape or suffer some distortion, but it will not break. A tough material has some flexibility, for example mild steel, which will not break when struck repeatedly. Hence we are able to cold form support brackets and the like from a mild steel bar. Cast iron is a brittle material and if we attempt to bend or shape cast iron it will break.

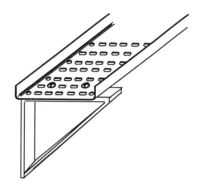

Figure 5.20    Support bracket made on site from mild steel

## Brittle

A material which, whilst it may be quite hard, does not withstand external forces particularly well. If brittle materials are bent or struck with a hammer they will usually break. Typical examples would be cast iron, glass and ceramic tiles.

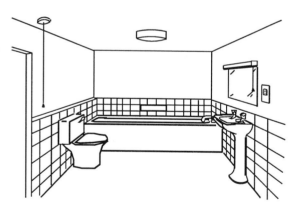

Figure 5.21    Tiles are brittle

## Elasticity

The elasticity of a material is its ability to return to its original size and shape after the application of an external force. PVC conduit has a degree of elasticity which is why bends made in the conduit during installation have a tendency to straighten out if they are not fixed in position. Of course the conduit will never return to its original shape and the most elastic materials are those like the rubber compounds used to manufacture tennis balls and rubber bands.

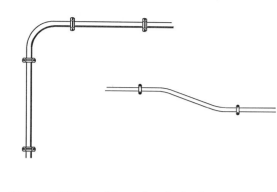

Figure 5.22    PVC conduit needs to be fixed in position or it will tend to straighten out.

## Ductility

A material which can be drawn out to form wires or strands is said to be ductile and these materials feature quite large in the manufacture of cables. Ductile materials include copper, used for conductors, and mild steel, as used for the armour of swa cables. Wrought iron is also a ductile material and this makes it suitable for the decorative shapes and structures made using wrought iron.

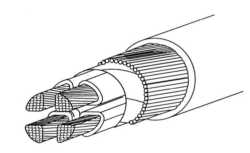

Figure 5.23    SWA cable

## Malleability

A malleable material is one which can be shaped, by pressing or beating, and will retain the new shape without breaking. Some materials are malleable when heated, such as wrought iron and mild steel, others are malleable when cold, such as lead.

## Ferrous metals

Ferrous metals are those which contain iron and typical examples are steels and cast iron.

## Non-ferrous metals

Non-ferrous metals are those which do not contain iron, typical examples being copper, aluminium and lead.

## Conductivity

Conductivity is related to a material's ability to conduct and there are two principal areas that are of concern to us, these being the ability to conduct electricity and heat.

### Electrical conductivity

Materials which freely conduct electricity are referred to as conductors and include materials such as gold, copper and aluminium. Materials which are poor conductors of electricity are referred to as insulators and include materials such as PVC, some oils, glass and porcelain.

### Heat conductivity

There are some materials which readily conduct heat, such as copper, brass and steel which are used to maximise this property in the manufacture of equipment such as boilers and saucepans. Materials which do not conduct heat well are referred to as thermal insulators and typically include glass, as in fibreglass, wood, some plastics such as polystyrene and some organic materials such as rockwool.

# Building materials and fixing components

Electrical installation work is carried out in buildings which will vary greatly in age, construction and use. As a result there will be a considerable variety of materials and construction methods which we will encounter. These in turn will require a variety of fixing methods and we need to select the most appropriate fixing method for the material, the means of construction and the equipment which we are fixing.

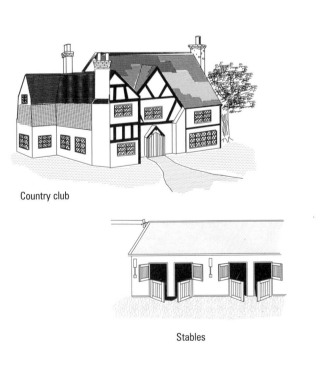

Country club

Stables

*Figure 5.24    Buildings will vary greatly in age, construction and use.*

*Remember*

The method of fixing used must be suitable for

- the material
- the method of construction
- the equipment which is to be installed
- the environmental conditions which are likely to prevail

and

- to be able to provide the necessary support without damage to the material, construction or the equipment

The types of materials used were considered in the Starting Work book in the Foundation Course series of these books, but a brief recap is appropriate here.

## Brickwork

Common or facing bricks are made from clay. They are easy to drill and have good fixing properties. Engineering bricks are harder and fairly difficult to drill. The most common methods of fixing are screw fixings using plastic plugs and masonry nails. The common additional tools which are used to create these fixings include electric and battery drills and rotary hammer drills. Masonry nails may be either hand driven or installed using a cartridge fixing tool.

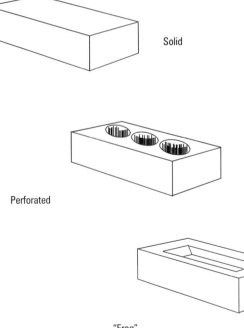

Solid

Perforated

"Frog"

*Figure 5.25    Brick types*

## Concrete

Concrete blocks are made from cement and aggregate which are usually in one of two categories, "lightweight" which are easy to drill and have fair fixing properties and "dense" which are fairly easy to drill and have good fixing properties. The most common fixing methods used for concrete blocks are screw fixings using plastic plugs, anchor bolts which use a number of different techniques, resin and stud anchors. The common additional tools which are used to create these fixings include electric and battery drills and rotary hammer drills. Masonry nails may be either hand driven or installed using a cartridge fixing tool.

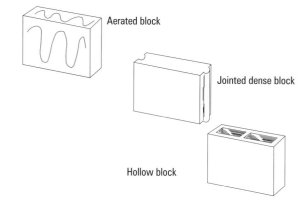

Aerated block

Jointed dense block

Hollow block

*Figure 5.26     Block types*

## Metal

Metals play a significant role in the construction of many larger buildings with steel being used to construct load-bearing frames for roof, walls and floors. Lightweight metal frameworks are used for ceiling grids, floor supports and structural components. Alloys are often used for the cladding and finishes of metal panels. The fixings used will vary dependant on the type of metal involved, its thickness, structural purpose and strength. Snap on fixings and girder clips are often used for structural steel beams. Although cartridge fired fixings, welding and other techniques may be used advice should always be obtained from the structural engineer before such fixings are made. The strength of the structure and the steel supports in particular can be seriously weakened by such fixings and qualified advice is necessary before such methods are used. Lighter steel panels and structures may be fixed to using self tapping screws, pop rivets, nuts and bolts and drilling and tapping.

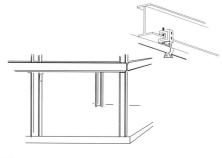

*Figure 5.27*

## Lining materials

These include plasterboard, lath and plaster, hardboard and fibre and particle board which are often used for lining the internal walls and ceilings of buildings. There are construction techniques which use bonded layers of plasterboard to create internal partition walls of minimal thickness. However these materials are most often used to provide a basis for decorative finishes and do not, in themselves, have a great physical strength. It is therefore not always easy to get a good fixing into these materials and so special fixings have been developed to overcome these problems. These fixings include spring and gravity toggle fasteners, collapsible sleeves (interset), cavity wall anchors and "easydrive" type fixings. Dry lining boxes are often used for accessories as these provide a means of fixing the boxes by the use of expanding lugs, spreading the load and providing a secure fixing without further drilling.

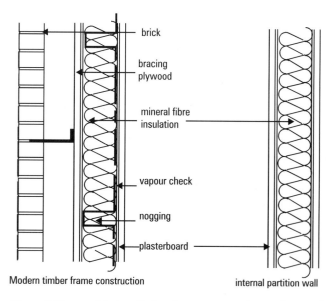

brick

bracing plywood

mineral fibre insulation

vapour check

nogging

plasterboard

Modern timber frame construction

internal partition wall

*Figure 5.28     Modern timber frame construction and internal partition wall*

## Wood, plywood and chipboard

Timber plays a significant part in the construction process particularly in dwellings. The use of timber joists and roof trusses continues to be the main stay of the domestic dwelling with many properties being constructed with timber frames, the exterior often clad with brick and the interior lined with plasterboard. Chipboard has, to a large extent, replaced the tongue and groove flooring in properties as the normal medium. Traditional timber floors are being installed as a decorative feature rather than a construction requirement. Plywood, thanks to modern treatments and adhesives, is manufactured to suit a variety of applications including use outdoors in quite hostile conditions with the use of marine ply. Fixings in timber are still largely made with the traditional wood screws and nails. Screws with special threads, such as twin thread and coarse chipboard threads are available to provide a screw fixing with maximum grip and minimum damage for all types of composite wood panels.

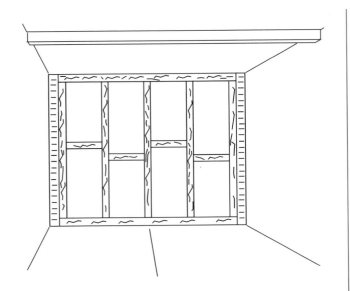

*Figure 5.29      Stud walling, wooden joist and flooring.*

# Part 4

## Effect of structure on site surveys

Having looked at the construction and materials used in the formation of buildings we need to be aware of how these can affect our site procedures during a survey or construction work. There are, as we have already discussed, responsibilities on both the employer and the employee with regard to health and safety in the workplace. There are some areas which introduce particular hazards and we shall consider some of those here.

## Roof work

Almost 20% of fatalities in construction work accidents involve people carrying out roof work. The majority of these are those in roofing trades but a number are those carrying out cleaning, repairs and maintenance to roof mounted equipment. It is important to remember that suitable facilities must be provided for access to flat and sloping roofs.

However some deaths and injuries are the result of falls through sheet roofing materials, such as plastic, asbestos, cement, fibreglass and corrugated iron. Sheet roofing materials tend to become brittle with age and iron rusts, and without suitable access equipment there is a real risk of falling through these sheeting materials.

Roofs should be accessed using purpose-made roof ladders or crawling boards to spread the load and allow suitable access up and down the roof. If the work is likely to last a short time then working from ladders and crawling boards may be OK but for longer duration suitable platforms, handrails and the like should be installed. Further details on the requirements may be found in Health and Safety in Construction Work, HS(G)150.

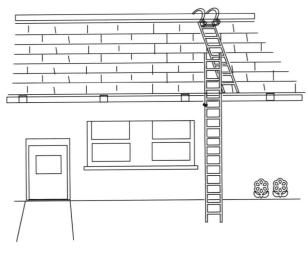

*Figure 5.30      Roofing ladder*

## Excavations and underground services

Care needs to be taken where work involves access to and use of excavations, which is often the case with services laid to new buildings and outside equipment such as street furniture. Working in or around excavations and trenches involves four main areas of danger.

### People and vehicles falling into excavations

People and vehicles may be prone to falling into excavations which are not properly guarded. Excavations which are more than 2 m deep should be protected by substantial barriers to prevent such incidents occurring and all excavations in public places must be suitably guarded.

## Collapse of trench walls or roof

The sides, and roof where one exists, of an excavation need to be suitably protected to prevent collapse. This may be achieved by "battering" the sides where appropriate. This involves cutting back the edges of the trench at a suitable angle and over sufficient distance to prevent collapse. If this cannot be achieved then suitable additional supports need to be installed to prevent collapse of the sides and roof. A range of proprietary equipment is available to allow this to be undertaken including trench boxes, timber and props and hydraulic side walling. Some of these can be installed so that persons do not need to enter the excavation to put them in place.

*Figure 5.31    Supported trench*

## Undermining other structures

Excavations carried out adjacent to existing structures such as buildings and walls may seriously affect the stability of the structure. Deep excavations carried out adjacent to buildings can cause the building, or part of the building to collapse with serious consequences. Even shallow excavations alongside garden or boundary walls, which may have quite shallow foundations, could cause the wall to collapse. This may not be immediate and could result in a serious accident whilst work is being carried out in the excavation.

*Figure 5.32    Excavation close to a wall could cause the wall to collapse.*

## Material falling into excavations

Plant, equipment, materials, vehicles and spoil should not be stored close to the sides of the excavation. The extra loading may cause the sides of the excavation to collapse and the stored items may fall onto persons in the excavation. A barrier should be installed to prevent vehicles passing sufficiently close to the edges and so cause the sides of the excavation to collapse.

*Figure 5.33    The sides of the excavation could collapse and fall onto persons working below.*

*Remember*
When working in an excavation hard hats need to be worn as you will effectively be working "below other activities" in much the same way as you would be on any construction project involving work above ground level.

## Underground services

The risks from underground services range from damage and injury as a result of striking services whilst excavating, to damaging services due to work carried out once the excavation is constructed. To avoid the damage whilst excavating, details of the routes and positions of any underground services should be obtained. Where these are not complete or available then the services may be traced using a variety of specialised techniques to provide information as to their location. In such circumstances excavation should proceed with particular care.

Once the excavation is complete the work carried out should have regard for any services which pass through the excavation. Where service pipes such as sewers, drains, gas and water pass through there may be a need for additional support to prevent damage through accidental contact, as shown in Figure 5.31. This would include equipment being placed or dropped onto the services and people standing on them. We must also have due regard for the effects of the work we are carrying out. Care should be taken when using heat or undertaking cutting or chemical processes in excavations.

# Confined spaces

Finally we shall consider working in confined spaces which attracts a number of hazards. Being in close proximity to moving machinery, electrical equipment, heat and cold sources such as steam pipes and refrigeration pipes creates a particular hazard for operatives in confined spaces. There is little room to avoid, or indeed to get away from, these hazards and where entrances are small and awkward the time taken to escape may be considerable.

A further complication may be the condition of the air available within the confined space. This may be made unbreathable by contamination from gas or fumes and, where there is inadequate natural ventilation, the quality of air may fall with time as work proceeds. There is also a risk of fire or explosion within the confined space as explosive gas and vapour or flammable dust and fibres may build up rapidly in a confined area.

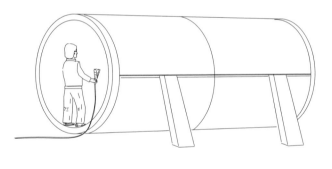

*Figure 5.34    The quality of air may fall in a confined working space.*

Expert advice should always be sought when work is to be carried out in such areas and appropriate protective clothing, equipment and, where necessary, breathing apparatus should always be worn. Work should not be carried out alone in such areas and someone should always be outside, at the access to the confined area, to communicate with those inside and to raise the alarm and take charge of the rescue procedures in the event of an accident. It is vital that all those involved in the work are aware of the procedures that need to be followed in the event of an emergency occurring.

Further information on all of these activities can be found in HS(R)150, Health and Safety in Construction.

It is important to establish the type of building construction, the external influences, purpose of the building and any special considerations that may affect a survey or electrical installation.

The influences on a survey or electrical installation include:

Building:          type of material, size and shape, movement and degree of flexibility in layout

Environment:     temperature, humidity, water, foreign bodies, corrosion, hazardous areas, flora and mould, livestock and vermin, mechanical impact, vibration, lightning strikes and electromagnetic effects

The IP code provides guidance on the degree of protection offered against ingress of solid bodies and water.

The IK code provides guidance on the protection offered against mechanical impact.

The materials used may possess certain properties and strengths, including hardness, toughness, elasticity, brittleness, ductility, malleability and conductivity.

Typical materials used will require a variety of fixing methods to cater for their particular properties. The materials include brickwork, concrete, metals, lining materials and timber.

Additional care needs to be taken when working in certain conditions such as on roofs, in and around excavations and in confined spaces.

## Self-assessment short answer questions

1. Within an industrial installation a number of machines are to be connected to a fixed wiring system of trunking and conduit. Explain briefly, with the aid of sketches, three methods of connection between machine and wall mounted isolator-starter that will prevent the transmission of vibration.

2. A supply is required to a piece of equipment some distance from a distribution board. For part of the route it will pass through underfloor ducts and part is to be suspended at high level. Select a wiring system that will, in your opinion, be the most suitable for the whole route giving the reasons for your choice.

3. You are to carry out an installation using PVC conduit and single core, PVC insulated cables with copper conductors. State the factors that you must consider to determine if this system would be suitable.

4. Identify suitable materials for the following applications
   (a) support brackets for trunking supports
   (b) electrical conductors
   (c) electrical insulators
   (d) heat conductors
   (e) heat insulators

5. Identify the hazards which may be present when working in a confined space.

# 6

# Assessing and Maintaining Quality

Before you start work on this chapter, complete the exercise below to ensure that you remember what you learned earlier.

The type of building materials used in the construction of a building may have an effect on the type of _____ and tools used to carry out a survey or the construction of an electrical installation.

It is necessary to establish the use and occupancy of the building and whether there are environmental conditions which will affect the choice of _____ or _____ methods.

Environmental conditions include temperature, humidity, water, foreign bodies, _____ substances, flora, livestock, vermin, mechanical _____, vibration, _____ radiation and _____ effects.

The materials used will have properties which make them suited to particular tasks. These physical properties include _____, hardness, _____, elasticity, ductility, malleability and conductivity.

The types of fixings and materials used need to be _____ with the structure, _____ and equipment being fixed without causing damage.

Additional care needs to be taken when working on _____, in trenches, _____ spaces and in hazardous areas.

## On completion of this chapter you should be able to:

◆ describe assessment methods to survey existing installations
◆ state the importance and means of effectively establishing appropriate technical information and recognising legal and regulatory requirements
◆ state means of maintaining quality standards
◆ explain the actions to be taken and the possible implications of variances from the specified standards
◆ explain the importance of effective communication in quality control

# Part 1

In the book "Stage One Design" in this series we have considered the requirements for electrical installations encompassing the

• Standards to which installations should comply,
• Regulations which cover the working practice and the completed installation
• methods of testing electrical installations.

Having given consideration to these requirements during the design of the installation, we must look at how we can demonstrate that the installation meets the design criteria throughout the construction. When the installation is energised and put into service it is important that we can demonstrate that it complies with the design criteria, any associated standards and that it is safe to use.

# Quality

The term quality is generally used to identify the standard of a particular item or service. Many companies are now Quality Assured to ISO 9000 or an equivalent standard. This, in simple terms, means that the company has systems in place to ensure that the work they carry out follows a controlled procedure on each occasion. This is to ensure that the standard of the product or service is consistent at all times.

*Figure 6.1*        *Quality symbol such as used for ISO 9000*

Companies which are Quality Assured are then subject to both internal and external surveillance to ensure that appropriate systems are in place and that they are followed. If an event occurs which shows a failing in the system a revised procedure would be required to prevent a recurrence. In many ways this is a good practice and most of us, as individuals, develop our own quality scheme based on personal experience. We change what we do on the basis of good and bad experiences, the quality scheme ensures that everyone involved in an activity is aware of the changes and implements them.

So quality can relate to the standard of a product or service and a system for ensuring consistency. It is important to remember that the introduction of a quality scheme does not necessarily mean an improvement in product quality, but it does ensure consistency of the existing product standard.

How does this affect our activities in electrical installation? As we have already discussed, we need to ensure the installation is installed to, and complies with, the appropriate standards. Once the installation is complete, we also need to inspect and test to ensure that it meets those standards and is safe to be put into service. This can be considered as the final quality check of the product.

**INSTALLATION SCHE**

Contractor

Test date

Signature

Method of Protection Against Indirect Contact

Items Not Tested

Distribution Board   Location of distribution board      Supply to distribution

| Description of work completed | | | | | | |
| --- | --- | --- | --- | --- | --- | --- |
| | | | Overcurrent Device | | | |
| Circuit | Circuit Description | Inst. Ref. Meth. | Type | Rating A | Short-circuit cap. kA | |
| | | | | | | |
| | | | | | | |
| | | | | | | |
| | | | | | | |
| | | | | | | |
| | | | | | | |
| | | | | | | |

Figure 6.2      Schedule of test results

Whilst we have checked the quality of the installation, we need to have a procedure in place to record what we have done so that we can demonstrate that the installation complies. This involves the keeping of records related to the design, both our own and others, the installation process, which may include installation and record drawings, and the inspection, testing and verification of the installation.

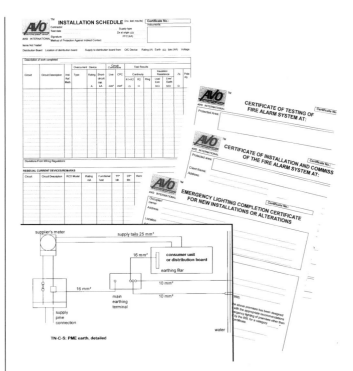

Figure 6.3      Records must be kept

In addition we may be asked to assess an existing electrical installation for

- an alteration or addition
- change of use
- a particular licensing application
- compliance with the requirements of BS 7671 and whether it is safe for continued use

Following any such assessment we need to provide documentation to show what we inspected and tested and, equally as important, what we did not inspect or test.

Not only do such documents demonstrate the quality of our installation, or the standard of an existing installation, they may also provide evidence that the requirements of the Electricity at Work Act have been met. In the unfortunate event of an incident occurring on an installation, on which you have worked, your documentation is the only evidence you have of the work you carried out, and the tests you have undertaken to ensure that the installation was safe to be put into service.

Figure 6.4

The requirement to maintain these records applies to all aspects of the installation work from design to commissioning and placing into service. Before we consider the final documentation required we need to review the requirements for the records collected during the design and construction phases of the installation. In the first instance we shall consider the requirements for a new installation. This will also apply to additions and alterations to an existing installation as the requirements are the same for all electrical installation work.

# Design data

During the course of the design of the electrical installation we need to determine the requirements of the installation for its intended use. This will include the client's requirements, the loading of the equipment to be installed, the maximum load, the characteristics of the supply and any external influences which may affect the electrical installation. Upon this background of information we can progress the design and select wiring systems, control and protection devices, cable types and sizes working to the final design for the installation. We need to maintain a record of all the information that we use to design the installation. Any subsequent change to this detail may result in the design not being compliant with the appropriate standards.

The system used for keeping the design data on record needs to be established at the outset and rigorously maintained throughout the course of the installation. The information should always be date-stamped and the date of receipt recorded. Similarly any information you issue should be uniquely identified, usually by the use of a reference number, the status of the revision and the date of issue recorded for both the recipient and yourself.

During the course of the installation process a number of changes may occur to a particular aspect of the installation. On each occasion there could be some cost implication whether in material, labour or time taken to complete the project. In order to recover any costs incurred detailed records are vitally important.

The information used for the changes to the design and any calculations supporting the final solution should also be recorded, in order that the design can be substantiated at a later stage if necessary. Manufacturers' details confirming power requirements, starting currents, operational criteria and the like should be used for the design. They should also be included in operation and maintenance manuals which will be issued to the client on completion of the work.

# Installation details

There may be changes to be made to the installation once construction has begun and the details of the work involved in the changes must be recorded. The extent of the detail required is similar to that we need to record for the design modifications.

Once part of the installation is in place, and other trades, such as the decorative finishes, begin their work, the actual location of the component parts of the electrical installation will be difficult to establish. Before this happens, we need to gather the necessary information in order to produce any record drawings necessary. The record drawings should be compiled as the work progresses. These drawings form part of the detail necessary to demonstrate that the installation was installed as it was designed.

*Figure 6.5*        *Electrical Installation and Minor Works forms*

Any testing carried out during the construction stage of the installation should be recorded. We need to record the following in order that the construction aspects can be confirmed:

- details of the circuits or cables being tested
- the results
- the date the testing was completed
- details of the instruments used, including the serial numbers
- detail and signature of the person who carried out the tests

# Certification

On completion of the installation we are required to issue, to the person ordering the work, our client, a certificate appropriate to the work we have carried out. The requirements for certification are defined in Part 7 of BS 7671: 1992 and IEE Guidance Note 3. Any certification issued should contain at least the information required in the standard forms set out in BS 7671 and be accompanied by a schedule of test results, detailing the design and test details of each circuit. Guidance Note 3 contains a pro-forma of a typical schedule of test results indicating the minimum information which should be provided. Electrical contractors are free to develop their own forms of certification or buy ready printed forms. A number of organisations within the electrical industry produce forms of certification for use by contractors. However it is advisable to check that, whichever source the certificates are purchased from, the final form provides all the information required by BS 7671.

# Part 2

There are a number of forms of certification and we need to be familiar with the purpose and compilation of each form. We shall begin by looking at the Electrical Installation Certificate.

## Electrical Installation Certificate

This certificate is used to certify new electrical installation work and may be used to certify a complete installation or alterations and additions to an existing installation. It should be issued by the contractor responsible for the construction of the electrical installation and a separate certificate should be issued for each distinct installation. A certificate must be issued to the person who ordered the work irrespective of whether they, or their client, have requested a certificate.

For the purpose of this text we shall be considering the minimum content of the Electrical Installation Certificate and it would be helpful to have a copy of BS 7671 and IEE Guidance Note 3 available for your reference.

The certificate should detail the following;

### Details of the client and the installation

The client is the person who ordered the work, your client, who may be the main contractor, the architect or principal electrical contractor. The client will not always be the owner or indeed the end user of the installation. The installation details should contain sufficient information to allow the installation to be clearly identified. The information should also state whether the installation is new, an alteration or addition and quite often more than one of these categories is applicable. New circuits may be installed along with additional and alterations to existing circuits when buildings are modernised or extended.

The extent of the installation which is covered by the certificate should be clearly identified. This information identifies the extent of the work for which you are responsible and should be completed in every case with a clear description. It is acceptable to identify either the extent of the work undertaken or any exclusions from the certificate whichever is the lesser. For example a complete rewire of a commercial property with the exception of the car park lighting would be better described by identifying the electrical installation at the address, and excluding the electrical installation and the control equipment for the car park lighting, in the extent section of the form.

### Design, construction and inspection and testing

The Electrical Installation Certificate should be issued by the person responsible for the construction of the installation. However, the same person may not be responsible for all aspects of the design, construction and inspection and testing of the installation. The certificate contains a declaration that all these aspects of the electrical installation are in accordance with the requirements of BS 7671. There are two options available:

1. Where all aspects of the design, construction and inspection and testing are carried out by the same person, there is the provision of a single declaration box for all three aspects and therefore requires just a single signature.

2. Where the responsibility for the design, construction and inspection and testing of the installation is not by the same person then each part must be declared separately. A separate section is included for each aspect and should be signed accordingly. The design of the installation may not be carried out by one person, for example where a consultant provides some aspect of the design, distribution circuits and the like, and the final circuit design is carried out by the contractor. In such circumstances there is provision for each individual to sign for their design, the extent of the design for which each is responsible is not detailed on the certificate. The design data produced by each party serves to provide this information, should it be required at a later date.

### Supply characteristics and earthing arrangements

In this section of the certificate we detail the information relevant to the supply for the installation, most of this detail we had to establish for the design of the installation. Here we confirm the details as they are on site. This includes;

- the type of system, TN-S, TN-C-S or TT being the most common types

- number and type of live conductors. Remember this includes neutral conductors but not earthing conductors

- nature of the supply This includes the nominal supply voltage, the frequency, external earth fault loop impedance and the prospective fault current.

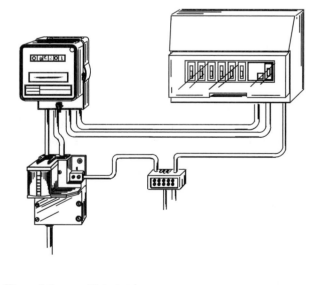

*Figure 6.6     Main intake*

## Particulars of the installation at the origin

Here we provide the information particular to the main intake of the installation. We need to include information on the means of earthing, details of the earth electrode where one is fitted, the earthing and main bonding conductors, the maximum demand and the main switch or circuit breaker. The details should include current ratings, BS type, location, csa and the like.

## Comments on the existing installation

When the certificate is used to certify alterations and additions to an existing installation then it is important to record any defects in the existing installation which do not affect the safety of the installation. Defects which would result in a reduced level of safety for the new installation must be rectified before the new installation is placed into service.

When the installation is new, or there are no defects found in the existing installation, then this section should be completed as "none observed".

## Next inspection

Once the installation is completed, the electrical installation certificate should provide information on the date at which the next inspection and test is due. IEE Guidance Note 3 gives some guidance on the maximum period between inspection and test for installations in a variety of types of premises. These values are the maximum periods and shorter periods may be appropriate dependant on the use, location and environmental influences which affect the installation. If a shorter period is recommended then we must be able to justify why this is required.

# Schedules

Included with the certificate should be the schedules of items inspected and tested and the schedule of test results.

## Items inspected and tested

The schedule items relate to the items which need to be inspected and tested in accordance with BS 7671. However, not all the items required by BS 7671 occur on every installation. Those which do not apply should be recorded as not applicable (N/A). Common examples of these would be Obstacles and Placing out of reach. These are methods of providing direct contact protection and are only used in special circumstances. It is important to record all the items which have been inspected and tested and to ensure that these cover all the requirements of BS 7671.

AVO INTERNATIONAL

**INSPECTION SCHEDULE**

Certificate No.:

**Items Inspected**

1. Connection of Conductors
2. Identification of Conductors
3. Routing of cables in safe zones or protected against mechanical damage
4. Selection of conductors for current/voltage drop
5. Connection of single pole devices for protection or switching in phase conductors only
6. Correct connection of socket-outlets and lamp holders
7. Presence of fire barriers and protection against thermal effects
8. Method of protection against electric shock
   a) Protection against both direct and indirect contact
      - SELV (note 1)
      - Limitation of discharge of energy
   b) Protection against direct contact (note 2)
      - Insulation of live parts
      - Barrier or enclosure
      - Obstacles (note 3)
      - Placing out of reach (note 4)
      - PELV
   c) Protection against indirect contact
      (i) EEBAD including:
      - Presence of earthing conductors
      - Presence of protective conductors
      - Presence of main equipotential bonding conductors
      - Presence of supplementary equipotential bonding conductors

      (ii) Use of Class II equipment or equivalent insulation (note 5)

      (iii) Non-conducting location (note 6): Absense of protective conductors

      (iv) Earth-free local equipotential bonding (note 7): Presence of earth-free equipotential bonding conductors

      (v) Electrical separation (note 8)
9. Prevention of mutual detrimental influence
   - Proximity of non-electrical services and influences
   - Separation of Band 1 and Band 2 circuits or Band 2 insulation used
   - Separation of fire alarm and emergency lighting cables
10. Presence of appropriate devices for isolating and switching correctly located
11. Presence of undervoltage protective devices where appropriate
12. Choice and setting of protective and monitoring devices (for protection against indirect contact and/or overcurrent)
    - Residual current devices
    - Overcurrent devices
13. Labelling of protective devices, switches and terminals
14. Selection of equipment and protective measures appropriate to external influences
15. Adequacy of access to switchgear and other equipment
16. Presence of danger notices and other warnings
17. Presence of diagrams, instructions and necessary information
18. Erection methods
19. Requirements of special locations

Inspected by ............ Date(s) Inspected ............

Notes
1. SELV An extra-low voltage system which is electronically separate from earth and from other systems. The particular requirements of the Regulations must be checked (see Regulation 411-02)
2. Method of protection against direct contact - will include measurement of distances where appropriate
5. Use of Class II equipment - infrequently adopted and only when the installation is to be supervised (see Regulation 413-03)
6. Non-conducting locations - not applicable in domestic premises and requiring special precautions (see Regulation 413-04)

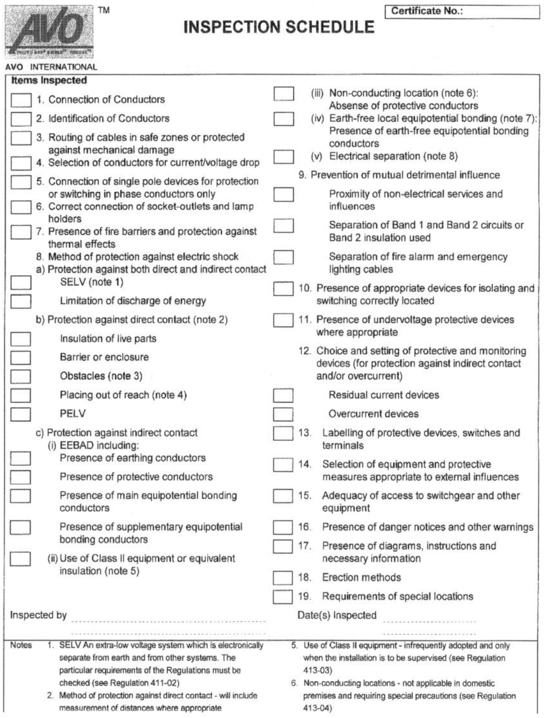

*Figure 6.7 Inspection schedule form*

## Schedule of test results

AVO INTERNATIONAL

**INSTALLATION SCHEDULE** (inc. test results)

Certificate No.:

Instruments:

Contractor ............ Supply type ............

Test date ............ Ze at origin (Ω) ............

Signature ............ PFC (kA) ............

Method of Protection Against Indirect Contact ............

Items Not Tested ............

Distribution Board | Location of distribution board | Supply to distribution board from | O/C Device | Rating (A) | Earth (Ω) | Ipsc (kA) | V

Description of work completed ............

| Circuit | Circuit Description | Inst. Ref. Meth. | Overcurrent Device | | | Circuit Conductors | | Test Results | | | | | |
|---|---|---|---|---|---|---|---|---|---|---|---|---|---|
| | | | Type | Rating | Short-circuit cap. | Live | CPC | Continuity | | | Insulation Resistance | | Zs |
| | | | | | | | | R1+R2 | R2 | Ring | Live/ Live | Live/ Earth | |
| | | | | A | kA | mm² | mm² | Ω | Ω | | MΩ | MΩ | Ω |

Deviations From Wiring Regulations:

**RESIDUAL CURRENT DEVICES/REMARKS**

| Circuit | Circuit Description | RCD Model | Rating mA | Functional Test | TT* Idn | OT* Idn | Remarks |
|---|---|---|---|---|---|---|---|

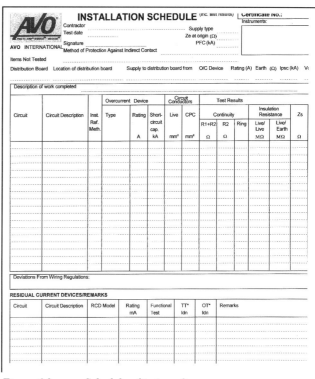

*Figure 6.8 Schedule of test results*

This is where we record the information regarding each individual circuit, detailing the design criteria and test results. Larger installations may well need to have a number of schedules of test results produced, and providing each schedule clearly identifies the part of the installation to which it applies, this is an acceptable way of providing this information.

The first information we need to provide is that for the distribution board and we need to include where it is located and what identification reference it has been given. We also need to identify where it is supplied from, the nominal voltage, number of phases and the type and rating of the protective device for the distribution circuit.

Where the distribution board is remote from the origin of the installation we also need to record the values of earth fault loop impedance and prospective fault current measured at the distribution board.

Where the distribution board is located at the main intake position this information will be as entered on the details of the installation at the origin and so will not need to be entered again here. Finally we need to record details of the test instruments used including the make and serial number for reference purposes.

It is generally easier to consider the test schedule in two parts, the first related to the circuit design details and the second to the test results. As an example, let's consider a single circuit supplied from a distribution board. The design aspects of the circuit are those which relate to the physical construction of the circuit including details such as the:

- circuit number and phase
- circuit description, which should be as descriptive as possible in order to distinguish the circuit from any other (lights, lights, sockets, sockets and so on are not suitable descriptions)
- type, rating and short circuit rating of the protective device, for example BSEN 60898, type "B", 16 A, 9 kA
- reference method for the installation of the circuit, as detailed in BS 7671
- type of wiring, for example PVC, PVC
- size of the live and protective conductors for the circuit.

This information should have been established at the design stage of the installation and would be confirmed on site.

The "second" part of the schedule of test results contains the results of the actual tests undertaken on the installation. We shall need to use these results to ensure that the installation complies with the design and the requirements of BS 7671.

The tests required by BS 7671 are covered in the Stage One Design book in this series and so we shall not be covering the test requirements again here.

The test results which need to be recorded are normally laid out in the sequence in which the tests are carried out. The sequence is detailed in BS 7671 and the majority of test schedules include those tests which are carried out for all installations, any specialist installations, or parts of installations, are usually separately certified by the specialist responsible for carrying out the tests.

So we need to record:

## Continuity

We would normally record this as $R_1 + R_2$, this being the most effective method of measuring the continuity of circuit protective conductors. Where the wander lead method of measurement is used we would need to record the $R_2$ value and clearly indicate that this is the value recorded.

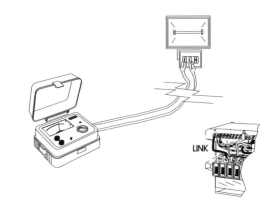

*Figure 6.9      Continuity test*

We should also record the ring circuit continuity values for any ring circuits. Where this section is not applicable as the circuit is not a ring circuit then we should enter N/A . Some schedules contain columns for recording all three values, $R_1$, $R_n$ and $R_2$ and as these need to be taken on site to confirm ring circuit continuity it is advisable to record the values on the schedule.

## Insulation resistance

We need to record the insulation resistance and ideally the full range of test values should be recorded for each circuit, that is phase to phase, phase to neutral, phase to earth and neutral to earth. Once again where the p-p value is not applicable, on single phase circuits, we should enter N/A.

## Polarity

We need to indicate that the polarity has been tested and is correct. The circuit should not be put into use if the polarity is incorrect. This is normally indicated by the means of a tick against each circuit once the polarity has been both tested and checked to be correct.

The above tests are carried out **BEFORE** the circuit is energised to allow the remainder of the tests to be completed in the knowledge that the installation is safe to energise. This is important for the safety of the people carrying out the testing and anyone in the vicinity at the time the tests are carried out.

---

*Remember*

The $R_1 + R_2$ test for continuity provides a positive indication for the polarity of the circuit. Once the circuit has been energised we need to check that the polarity is correct, this is normally done when the live tests are carried out. The appropriate boxes on the schedule of test results can only be ticked when the polarity has been both tested before **and** checked after the circuit is energised.

---

## Earth loop impedance

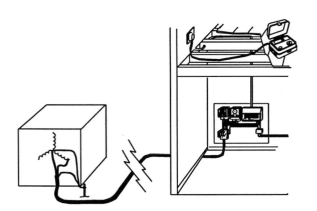

*Figure 6.10     Earth fault loop impedance test*

We need to test and record the maximum earth loop impedance for the circuit, and this is generally the value measured at the furthest point from the supply on each circuit. Measuring Zs at each outlet will determine the maximum value for each circuit and it is the maximum value which needs to be recorded.

## Operation of RCDs

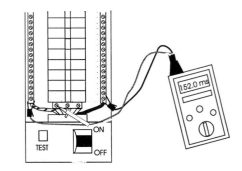

*Figure 6.11      RCD test*

There are two values which may need to be recorded here, the first being the operating time for the RCD at the rated current, so the disconnection time measured, for say a 30 mA RCD at 30 mA, would be recorded. The second is the operating time for an RCD of 30 mA or less which is installed to provide supplementary protection against direct contact. 30 mA RCDs installed to provide protection for socket outlets likely to supply portable equipment for use outdoors would be a typical application. In these cases the operating time of the RCD when tested at 5× the rated tripping current needs to be recorded.

---

*Remember*

Many RCD test instruments have the facility to carry out the test in both the positive and negative half cycles of the supply. The tests should be carried out in both half cycles and the higher of the two values obtained should be recorded. This represents the most onerous condition for the circuit. The results are recorded in milliseconds.

---

The test results obtained should be verified against the requirements of BS 7671 and the original design parameters in order to ensure that the installation complies with those requirements and that it is safe to be put into service.

Having considered the Electrical Installation Certificate we will now look at the Minor Electrical Installation Works Certificate.

# Part 3

## Minor Electrical Installation Works Certificate

**MINOR ELECTRICAL INSTALLATION WORKS CERTIFICATE**
(REQUIREMENTS FOR ELECTRICAL INSTALLATIONS - BS7671 [IEE WIRING REGULATIONS])
To be used only for minor electrical work which does not include the provision of a new circuit.

Certificate No.:

AVO INTERNATIONAL

**PART 1: DESCRIPTION OF MINOR WORKS**

1. Description of the minor works
2. Location/Address

3. Date minor works completed
4. Details of departures, if any, from BS7671: 1992 (as amended)

**PART 2: INSTALLATION DETAILS**

1. System earthing arrangements (where known)
2. Method of protection against indirect contact
3. Protective device for the modified circuit          Rating          A
4. Comments on existing installation, including adequacy of earthing and bonding arrangements : (see Regulation 130-09)

*Figure 6.12      Minor works form*

This certificate is intended to certify modifications to a single circuit only, where a new protective device is NOT installed, so it must not be used to certify a new circuit. The Minor Electrical Installation Works Certificate has some similarities with the Electrical Installation Certificate, although by the nature of the certificate the content is somewhat reduced, occupying a single page.

Part One of the Minor Electrical Installation Works Certificate is where we record the details of the location of the work and the extent of the work undertaken. We also need to record the date the work was completed and any departures from BS 7671.

Part two of the certificate is used to record the details of installation which includes the type of earthing system, the method of protection against indirect contact, the type and rating of the existing protective device for the modified circuit and any comments on the existing installation.

Part three of the certificate is used to record the test results for the circuit. These are those tests essential to ensure the circuit is safe to put back into service. These include earth continuity, insulation resistance, earth fault loop impedance, polarity and RCD operation where one is fitted.

Part four of the certificate is the declaration where the person responsible for the works declares that the work does not impair the safety of the existing installation. The signatory also declares that the work carried out has been designed, constructed, inspected and tested in accordance with BS 7671 and that the work complies with the requirements of that standard.

The Minor Electrical Installation Works Certificate is designed for use under particular conditions and the extent of the information provided on the certificate reflects that. Where minor works are undertaken, the original existing installation should have an Electrical Installation Certificate and the Minor Electrical Installation Works Certificate is supplementary to that certificate.

Under the specific circumstances covering the issue of the Minor Electrical Installation Works Certificate we are verifying the suitability and safety of the modified circuit to be put into service. It is important to ensure that the existing installation

- is suitable for the intended work,
- will not have its safety compromised on completion of the proposed work

and

- the circuit we modify is compliant with the requirements of BS 7671.

This sometimes causes confusion as to the extent of the work necessary to ensure compliance with BS 7671.

As an example let us consider the standard domestic installation, where we are to modify an existing power circuit supplying the ground floor socket outlets. Before we begin the work, we need to establish the means of protection against indirect contact which is generally EEBAD (Earth Equipotential Bonding and Automatic Disconnection). Assuming this to be the case, we must ascertain that the existing installation has the necessary main equipotential bonding in place to provide that protection. If this is not in place or is not adequate then the bonding should be brought up to the required standard before the circuit is modified and put back into service. Similarly we need to consider whether any of the sockets on that circuit are likely to be used to supply portable electrical equipment outdoors. If so the circuit should be protected by an RCD rated at no more than 30 mA and if one is not already fitted then this would need to be installed before the circuit is modified and put into service.

*Remember*
The circuit modified and certified using a Minor Electrical Installation Works Certificate must meet the requirements of BS 7671 and not impair the safety of the existing installation.

Those parts of the existing installation which affect the compliance and safety of the modified circuit must also meet the requirements of BS 7671.

Other aspects of the existing installation which are observed that are not in accordance with BS 7671 and do not affect the compliance of the modified circuit should be brought to the client's attention and recorded on the Minor Electrical Installation Works Certificate.

You have an obligation under the Electricity at Work Regulations to ensure that the duty holder for the installation is informed, without delay, of any dangerous aspects of the installation which could affect the users of the installation.

The final form of certification we shall consider is the Periodic Inspection Report.

# The Periodic Inspection Report

This is different from the previous Certificates because it is not issued for electrical installation work. As the name implies, it is used to report on the condition of an existing electrical installation, one that has been placed into service. The report is used to record the compliance of an installation, which has been in use, with the current requirements of BS 7671.

*Remember*
A Periodic Inspection Report records the compliance of an existing installation with the CURRENT requirements of BS 7671, irrespective of the age of the installation or what regulations were effective at the time the installation was completed.

So why does an installation require a periodic report? Every electrical installation deteriorates with age and use, so it is important to ensure that the installation is safe for use and does not put users of the installation at risk. The frequency of the inspection is dependant upon a number of factors, such as the type of building and its use, the age of the installation, and the environmental conditions. IEE Guidance Note 3 provides some guidance on the maximum period to the first inspection and test of a new installation based upon these criteria. The frequencies between inspections are determined considering these factors and the condition of the installation following a period of use.

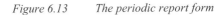

*Figure 6.13    The periodic report form*

The purpose of the periodic inspection is to establish, as far as is reasonably practicable, any factors which could impair the safety of the electrical installation and report on them. Chapter 73 of BS 7671 outlines the requirements for periodic inspection and testing of electrical installations.

So having established the need for periodic inspection and testing and the need to report on the findings, what of the format used for the report? Appendix 6 of BS 7671 and Part 5 of IEE Guidance Note 3 give standard forms for the periodic inspection and testing of an electrical installation. In addition to the standard forms we need to produce a schedule of items inspected and tested and a schedule of test results. The schedules and their compilation are generally identical to those for the Electrical Installation Certificate and so we shall not be considering the schedules again.

The first part of the form provides details of the client and their address, which may be different from that of the installation to be inspected and tested, and the reason for which the report is required. It is important to record the purpose for which the report is required. This may be for a license for a restaurant, hotel, public building or a particular use such as a public entertainment license. It may be required as part of a house sale survey or as a part of the duty holders insurance requirements or legal obligation.

The next section of the form is where we record the details of the installation. This requires information such as the occupier and the address of the installation.

*Remember*
The client may not be the occupier and they may not reside at the same address as the installation to which the report refers.

We must also record some information relative to the installation itself which includes the

- type of premises
- estimated age of the electrical installation
- evidence of any alterations or additions to the installation
- date the last inspection and report was carried out and whether there are records available

This information provides detail of the existing electrical installation upon which the periodic inspection and report is carried out and should always be filled in as accurately as possible.

The next box is where we detail the extent and limitations of the inspection and report and is one of the most important parts of the report. It is here that we have the opportunity to detail what was, or was not included in the inspection and any limitations placed upon us whilst we were carrying out the inspection.

These items should be agreed with the client before any work is undertaken so that both parties are aware of the extent of the work to be undertaken. The requirements of any third party, such as an insurance company or licensing officer must, of course, be taken into account. The overall extent of the work is however the subject of negotiation between you and the client, with you providing the technical advice as to the electrical and any additional third party requirements. Once these are agreed the proposed extent of the work and any foreseen limitations can be noted.

The true extent of the work and any limitations can only be recorded on the report once the work has been completed. If, for example, at the time of the inspection the user of the installation restricts your access, or parts of the installation cannot be isolated, then these items will need to be recorded. Then the recipient of the report will be accurately informed of the extent of the inspection and to what the report actually refers.

The next two sections can only be completed once the inspection and testing has been carried out and the condition of the installation determined. The date of the next inspection is, as we found earlier, dependant upon the condition of the installation, the environment, the type and use of the premises. The actual period to the next inspection is therefore based upon your findings and the period cannot be determined until the report has been completed. The second section is the declaration which, once again, cannot be completed until the inspection and testing has been carried out and the condition of the installation established.

The next section of the report deals with the aspects of the installation which need to be measured or ascertained and are similar to those at the origin of the installation recorded on the Electrical Installation Certificate and so we shall not be covering these again here. However there is one section which requires some discussion and clarification and that is the Observations and Recommendations section.

# Observations and recommendations

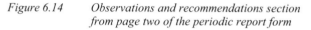

*Figure 6.14*    *Observations and recommendations section from page two of the periodic report form*

This is the section of the report where you make your observations on the installation and recommendations as to the action required. There are two important factors to bear in mind here.

## Observations

These should be observations based upon your findings and relate to those parts of the installation which do not meet the requirements of BS 7671. We are not required to record the regulation numbers as the report should be understood by a layperson. However we should be able to support each observation by a regulation with which the installation does not comply. Custom and practice should not be included in the observations, just because a particular aspect of the installation is not done as you would normally do it does not automatically mean it does not comply with the requirements of BS 7671.

## Recommendations

The recommendations are not instructions on how to correct the departures or information on what needs to be done. There is a standard code for recommendations and these are in the form of numbers 1, 2, 3 and 4 to indicate the action that needs to be taken.

In simple terms the recommendation codes are:

Code 1:
   Requires urgent attention: informs the recipient that potential danger exists and that this item requires immediate action.

Code 2:
   Requires improvement: informs the recipient that a deficiency exists in the installation which does not currently represent a danger but which requires improvement.

Code 3:
   Requires further investigation: informs the recipient that, within the agreed extent and limitation of the report the inspector was unable to come to a conclusion on this aspect of the installation. It is important to remember that the purpose of the inspection and report is not to carry out fault finding. If a circuit is found to be below the requirements for insulation resistance testing, for example, this should be recorded but the precise cause of the failing would not normally be found.

Code 4:
   Does not comply with BS 7671:1992 as amended: indicates that there is a departure from the requirements of BS 7671 but the users of the installation are not in any danger. A typical example would be where green only sleeving is used on the cpcs of a circuit whilst BS 7671 requires these to be coloured green and yellow.

It is important that the Observations and Recommendations are recorded in this way to provide standard information and recommendations to those receiving the reports.

---

*Remember*
There may be several alternative methods of achieving the necessary improvement required. The purpose of the report is to identify the need for improvement and the urgency with which this action needs to be undertaken. The precise remedy is a matter for discussion between the person responsible for the installation and the electrical contractor engaged to carry out the remedial work. Both will be using the information provided on the report to determine the most suitable and cost effective action to be taken to achieve a compliant installation.

---

The other sections we need to consider are the summary of the inspection and the schedule record.

The summary of the inspection is where we describe the overall condition of the installation. The comments made here should adequately describe the condition of the installation and help to substantiate your overall recommendations. If, for example, some aspects of the installation, whilst compliant at the time of the inspection, are showing signs of deterioration then this should be recorded here. It would be reasonable in such circumstances to recommend a shorter period to the next inspection as a result. Simple statements such as "poor condition" or "needs rewiring" do not adequately summarise the condition of the installation or provide sufficient information to substantiate your statement.

Finally we need to record the number of pages that accompany the report. These will include the schedules of items inspected and tested, the schedules of test results and any additional pages used to record the observations and recommendations or summary of the installation.

---

*Remember*
Always consider your observations, recommendations and summary in such a manner that if you were to receive the report you could provide a reasonable quotation for the remedial work, having never seen the installation. The information provided should therefore be sufficient to allow you to make a reasonable assessment of the actual condition of the installation.

---

*Points to remember* ◄ – – – – – – – – – – – –

Following any electrical installation work the appropriate form of certification should be issued to the person requesting the work.

These documents provide a record of the work for which you are responsible and details of the inspection and testing carried out to ensure that the installation was safe to put into service.

The certification also details any design departures from BS 7671 and identifies any noted deficiencies on the existing installation.

This information could prove to be invaluable should an incident occur on an installation upon which you have carried out work.

The two main forms of certification used for electrical installation work are the Electrical Installation Certificate and the Minor Electrical Installation Works Certificate.

The Minor Electrical Installation Works Certificate is only used for alterations and additions to a single circuit which does not include the installation of a new circuit.

The Periodic Inspection Report is the form used for reporting on the condition of an existing installation which has been energised and placed in service. It is not appropriate for certifying new installation work.

Periodic Inspection Reports may be required for a number of reasons and it is important to record on the report the purpose for which the report is carried out.

During a periodic inspection the inspector is considering the compliance of the electrical installation with the current requirements of BS 7671, irrespective of when the installation was originally constructed.

It is important to accurately record the extent and limitations of the report and to make clear and easily understood recommendations on the condition of the installation.

The observations made on the periodic inspection report should relate to the requirements of BS 7671 and not on custom and practice.

It should be possible to support the items which are recorded, with departures from specific regulations from BS 7671 although these do not need to be recorded on the report.

The inspector makes a recommendation on the period to the next inspection and test based upon the condition of the installation, its use and any external influences which may affect it.

## Self-assessment short answer questions

1. Identify the three principal forms of certification and briefly state the purpose for which each is used.

2. List three reasons for which a client may wish to have a periodic inspection of their installation carried out.

3. List the three signatories which are required on an electrical installation certificate.

4. In addition to the information detailed in the model forms in Part 7 of BS 7671 for the electrical installation certificate, what other information should be included with the certificate.

5. At what stage are the details of the extent and limitations recorded on the periodic inspection report.

# 7

# The Contract, Co-ordination and Control

Before you start work on this chapter, complete the exercise below to ensure that you remember what you learned earlier.

Before an electrical installation is placed into service it should be _____ and _____ to ensure that it meets the requirements of BS 7671 and any other appropriate _____.

Upon completion of the installation a _____ should be issued to the person requesting the work.

The _____ _____ certificate is used to certify new electrical installation work including _____ and _____ to existing installations.

The _____ _____ _____ is used to certify alterations and additions to a single circuit. It is not to be used where a ____ circuit is installed.

A periodic inspection report is used to report on the condition of an _____ installation. That is an installation which has been _____ and _____into service, it is not appropriate for the certification of new work.

Each form of certification should include a schedule of test results detailing the results of the _____ carried out on site.

## On completion of this chapter you should be able to:

◆ state the need for formal contracts and procedures
◆ identify the main points covered in a contract
◆ state the possible outcomes of failing to comply with the above
◆ state the relationship that exists between the parties involved in a contract
◆ list the organisations involved in maintaining industrial relations
◆ describe the means of developing planning schedules from specifications and contract requirements
◆ state the means of controlling contract progress
◆ state the importance of good customer relations

# Part 1

# Contract administration

## Contracts

In Chapter 1 we discussed briefly the use of contracts and we must now consider them in more depth, the first step being to establish exactly what is a contract. In its most basic form a contract is defined as an agreement between two parties. The agreement may be a verbal one, an agreement to meet a friend at a given time, even when made during a telephone conversation, could be classed as a form of contract. In the business world, however, it is more common to have written contracts which form a legal agreement between the parties involved covering all aspects of terms and conditions.

In order for a contract to exist certain conditions must be met. To illustrate these conditions we will consider the case of an electrical contractor being asked to carry out a simple wiring job. In this case the customer approaches the electrical contractor, asking if the contractor would be prepared to do the job and the cost of the work, in fact the contractor has been asked to quote or tender for work. The contractor provides his quotation for the work detailing the cost and time involved and the customer completes the process by agreeing to have the work completed based upon the quotation.

*Figure 7.1*

With small jobs this whole process may be completed verbally and no written contract produced. However should any dispute occur there would be some difficulty in resolving the matter without any written evidence, thus placing both parties at risk.

Similar principles apply to any job and the electrical contractor must be aware of the contractual position from the time that the enquiry is received from the customer.

So how is the contract made?

If we consider the contract between an electrical contractor and a customer, irrespective of the size of the job, the process is the same. First the customer, or client, has to ask the contractor to quote, or tender, for carrying out a job of work. The request should contain enough information to allow the contractor to determine the extent of the work and the time and material required to complete it.

Once the extent of the work has been established, the contractor must make an offer to carry out the work, in terms that are clear enough for the customer to understand. As far as our electrical contractor is concerned, the offer is an undertaking to carry out an electrical installation in return for a promise by the customer to pay the price for this work. This is the first stage in the forming of the contract.

Figure 7.2    The electrical contractor receives a request to quote for the job

Next there must be an unqualified acceptance of the electrical contractor's offer by the customer. It is this acceptance that completes the contract and until the offer has been accepted there is nothing under English law to legally bind the two parties. The contractor is free to withdraw the offer at any stage until it is accepted. When an acceptance is made, it must be to accept the tender on the terms put by the contractor. With larger contracts it is generally the customer who states the terms that are to be offered. The contractor must then tender on the basis of these requirements which are often referred to as the terms and conditions of tender. These define the terms by which the contractor must tender.

The term unqualified in this sense means that the offer is accepted exactly as stated in the wording of the tender. If the customer replied accepting the terms but wanted to, say, pay in instalments then, if this was not written in the original tender, the contractor could withdraw the offer. The contractor could, however, make another offer based on the new terms set by the customer.

Figure 7.3    The terms and conditions of the contract

The final requirement is that of consideration. The consideration in this case is the promise of the electrical contractor to carry out the installation in return for the promise by the customer to pay the price required. Under English law this price does not have to be a monetary one.

So those are the requirements for a contract and, as a result of numerous court cases since the 19th century, there are some basic rules laid down regarding the offer and acceptance of the contract. These are briefly as follows:

The person(s) making the offer must use language that is clear and concise and bring the terms of the offer to the person(s) to whom the offer is addressed.

Should they decide to accept, then an unqualified acceptance must be communicated back to the person(s) making the offer.

Until the communication of acceptance is received by the person(s) making the offer, the contract is not complete.

Figure 7.4    An acceptance of the offer

Having seen what is needed to form a contract we must also consider the common conditions that can prevent a contract being made or accepted. These are best summarised as follows:

**Lapse due to time:** In order to prevent any misinterpretation as to what is a reasonable time most offers carry a clause stating the length of time for which the offer will remain open. Any offer will automatically expire after this time period.

| MARCH | | | |
|---|---|---|---|
| 4 | 11 | 18 | 25 |
| 5 | 12 | 19 | 26 |
| 6 | 13 | 20 | 27 |
| 7 | 14 | 21 | 28 |
| 1 | 8 | 15 | 22 | 29 |
| 2 | 9 | 16 | 23 | 30 |
| 3 | 10 | 17 | 24 | 31 |

*Fig 7.5        A contract will expire after the time period in the offer has lapsed.*

**Withdrawal:** At any time until the offer is accepted the person(s) making an offer may withdraw their offer. This is irrespective of the length of time for which the offer was to be available. The legal requirement is that the contractor must notify the customer of the withdrawal and if this is not done the customer can still accept the offer within the time stated.

Note: In Scotland the promise to keep an offer is binding and so these conditions do not apply.

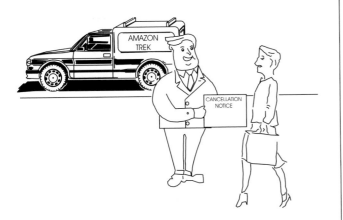

*Figure 7.6        The contractor must notify the customer of the withdrawal of the offer*

**Death of the contractor:** If the contractor should die before the offer is accepted the customer must be notified otherwise the contract could come into existence and would be perfectly valid.

*Figure 7.7        The customer must be notified if the contractor should die before the offer is accepted.*

**Rejection:** The contractor's offer can be rejected by the customer and no reason has to be given for the rejection. Once this has been done, the rejection is irrevocable and the customer cannot have a change of mind at a later date. The only recourse in such a case would be for the customer to ask the contractor to submit another offer. The contractor may choose not to quote again for the work. However, should a quotation be offered it need not be on the same terms or conditions as the original.

*Figure 7.8        The customer may reject the contractor's offer*

These are the main reasons for withdrawal from the contract process. Again there are a number of conditions that may exist that are not mentioned here.

*Remember*
If you are to be involved in any kind of contract, it is always advisable to seek professional advice and assistance before signing it.

Figure 7.9    *Legal action could result from unfulfilled terms of a contract*

Failure of a contractor or customer to fulfil the terms of a contract will almost always result in a legal action against the party concerned. This may prove costly to whichever party found to be in breach of the contract and so should be avoided. If a specification was issued, the contractor will have tendered the contract to this specification. The electrician carrying out the installation must ensure that the work carried out complies with the specification. Failure to do this could result in there being an action brought for breach of the contract.

Figure 7.10    *A court action could be brought for breach of contract*

# Part 2

## Parties associated with the contract process

In Chapter 1 we also considered the parties who could be involved in the design process. Similarly there may be a number of parties involved in the contractual process and, dependant upon the size and complexity of the work these could include;

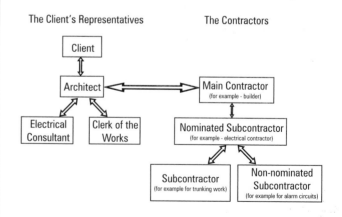

Figure 7.11

**Client:** the person who requests the work to be done. If the work is of any size, which could be from building a single dwelling to a multimillion pound construction, it is likely that the client will engage the services of an architect.

*Figure 7.12      The client*

**Architect:** appointed by the client for any work of reasonable size and may be responsible for the supervision and administration of the contract for the client. The architect acts as agent, negotiator and arbitrator for the client. Any changes to the original requirements are generally under the control of the architect. In practice the architect will often employ consultants for specialist areas.

*Figure 7.13      The architect*

**Consultant:** appointed by the client or architect for specialised advice and responsible for ensuring the specialist contracts, such as electrical and mechanical services, are technically correct and meet the requirements of the client.

*Figure 7.14      The consultant*

**Management contractor:** on larger jobs it is quite common to engage the services of a management contractor who performs such functions as letting main and subcontracts, programming and monitoring progress of work, alterations and additions, standard of construction and payment of trades.

**Main contractor:** the contractor who has been awarded the contract to produce the finished work in its entirety.

*Figure 7.15      The main contractor*

**Subcontractor:** These are usually selected by the main contractor who specifies the terms offered for their work. The subcontractor is not recognised by the architect and is not mentioned in the main contract.

## The electrical contractor

Most electrical contracts are tendered for against the specification drawn up by the electrical consultant, generally referred to as the "electrical works contract". The work is carried out under a separate contract to the main contractor or consultant and therefore the status of the electrical contractor will be that of a subcontractor. The effect of this is that the main contractor may be regarded as the customer and the electrical contractor will be bound by the terms of their contract with the main contractor.

*Figure 7.16      The electrical contractor*

# Contract co-ordination and administration

Where work involves a number of trades there are likely to be several subcontracts awarded. Each subcontractor will need to develop a planned schedule for the completion of their work. Each schedule will need to be co-ordinated with the others to ensure that the work is carried out within the time allowed and with the minimum of disruption to progress.

In order to achieve this it is common for the contractor responsible for the construction to produce an outline programme for the construction aspects of the work. Each contractor will produce a programme of their own aspects of the work which then have to be put together to form a working programme for the project.

*Figure 7.17     The contractor with a working programme*

The extent to which the individual programmes can be accommodated will depend on the time for the overall programme and the time available for each stage of the work. Compromise is almost always necessary with contractors working together to produce a workable programme. Because of this it is common to find trades sharing work areas and tight schedules. If one aspect of the work is delayed, it can have serious repercussions for the activities which follow, and it is for this reason that delays can result in considerable additional costs to the project.

The basis for the original programme is quite simple, the extent of construction or alteration to an existing structure is of prime importance, until a building exists there can be no work carried out within it. Likewise the building does not have to be, and generally cannot be, complete before work can commence on the services and internal construction. The approach needs to be logical and well thought out, for each activity there is some necessary preliminary work which must be in place before the activity can begin.

Take a simple example such as fixing a switch box to a wall. Before we can fix the box the wall must have been constructed and have set in the area where the box is located. The rest of the wall may not be complete but the area where the switch is to go

must be. The necessary tools must be on site to carry out the work, measuring tape, drill, drill bits, hammer and chisel.

*Figure 7.18     Before the box can be fitted the wall must be complete!*

We must also have the necessary material, switch box, wall plugs and screws all of which are fairly obvious. But we must also ensure that we can get to the location and there are not obstructions such as scaffolding, shuttering, plant or materials in the way. Equally important we must ensure that there is not other work scheduled in the area, such as laying flooring, spraying, overhead construction, welding and the like, which would prevent us from carrying out our work.

On a large project there are many aspects to be considered and in order to achieve the project completion on time the works must be integrated as effectively and efficiently as possible. Each contractor has an obligation to complete their activity within the programmed time and to the programme dates.

The interweaving of these activities is quite an art and, on large projects, professional work programmers are often employed to produce programmes, monitor and predict progress and advise on necessary changes and priorities.

## Planning

In this part we shall look briefly at the requirements for the planning, programming and scheduling of work. For a small installation we need to consider obtaining materials, arranging access, and planning when our next job is due to start.

We have considered all the activities necessary to produce a basic programme for our work using a simple logic sequence for carrying out the installation work.

Once the order is received we need to:
- order the material and organise labour for the start date
- deliver 1st fix materials and ensure the labour is on site for the start date
- deliver 2nd fix materials to the site just in advance of the second fix
- inspect and test the installation
- complete the documentation and certification for the work

During the site activity we need to monitor the progress of the work to ensure that any deadline dates are achieved. These are usually to allow "follow on" trades, such as plasterers, carpenters and decorators to complete their work on time.

This is an oversimplified procedure but the intention is to provide us with an idea of the approach required when we consider more complex projects. This type of planning is generally relayed to a bar chart in order that the project can be viewed as a whole and the stages of construction and completion checked on a regular basis. To keep this simple each activity must be complete before the next activity begins and we will then finish up with a chart similar to that shown in Figure 7.20.

Figure 7.19    A crowded working area

| Operation description | Day No. | | | | | | | | | | | | |
|---|---|---|---|---|---|---|---|---|---|---|---|---|---|
| | 1 | 2 | 3 | 4 | 5 | 6 | 7 | 8 | 9 | 10 | 11 | 12 | 13 |
| 1st fix installation | | ■ | ■ | ■ | ■ | | | | | | | | |
| 2nd fix installation | | | | | | ■ | ■ | ■ | ■ | ■ | | | |
| Inspection and test | | | | | | | | | | | ■ | | |
| Handover | | | | | | | | | | | | H/O | |
| | | | | | | | | | | | | | |
| Deliver 1st fix | ◆ | | | | | | | | | | | | |
| Deliver 2nd fix | | | | | ◆ | | | | | | | | |

Figure 7.20    Simple bar chart

---

*Try this*

You are asked to produce a programme for carrying out the rewire of some small domestic premises, for a refurbishment contract for a housing association. The work is likely to take one electrician seven days to complete from starting work to completion of the certification and documentation. Produce a Bar Chart showing how the work is to be scheduled to achieve this.

| Operation description | Day No. | | | | | | | | |
|---|---|---|---|---|---|---|---|---|---|
| | 1 | 2 | 3 | 4 | 5 | 6 | 7 | 8 | 9 |
| 1st fix installation | | | | | | | | | |
| 2nd fix installation | | | | | | | | | |
| Inspection and test | | | | | | | | | |
| Handover | | | | | | | | | |
| | | | | | | | | | |
| Deliver 1st fix | | | | | | | | | |
| Deliver 2nd fix | | | | | | | | | |

The bar chart shown in Figure 7.21 shows a more practical activity sequence and we can see that work is being carried out at more than one location at the same time. On larger projects this chart will become considerably more involved and many more activities will overlap.

| STAGES | WEEKS | | | | | | | | |
|---|---|---|---|---|---|---|---|---|---|
| | 1 | 2 | 3 | 4 | 5 | 6 | 7 | 8 | 9 |
| 1st fixing conduit (lighting) | ███ | ███ | ███ | | | | | | |
| 1st fixing conduit (power) | | ███ | ███ | ███ | | | | | |
| Wire conduit (lighting) | | | | | ███ | ███ | | | |
| Wire conduit (power) | | | | | ███ | | | | |
| 2nd fix lighting | | | | | | ███ | | | |
| 2nd fix power | | | | | | | ███ | | |
| Inspect and test | | | | | | | | | |

*Figure 7.21    Bar chart*

This is a typical working programme proposal which would be forwarded to the main contractor. Whilst this contains the detail relevant to our activities there are the other construction trades to be considered. For example the plumbing and air-conditioning first fix also needs to take place before the plastering and ceiling erection. In addition the builders' progress may be restricted due to curing times etc. and all these factors will have an effect on the final working programme.

# Part 3

## Critical path networks

As construction projects become increasingly more complex, the planning and the execution to ensure completion at a given time becomes more important. One way in which this is achieved is by the production of Critical Path Networks. The principal function of these networks is to identify the critical elements in the project which must be completed at a given time to ensure the overall project is completed to programme. These items may appear to be unrelated in terms of trade or process but form a vital link in the completion of the project.

With every contract there will be a Critical Path and the activities which make up the Critical Path must be carefully monitored to ensure completion on time. A simple Critical Path for a fairly basic operation where the activities are linear and the completion of each activity is vital to completion of the job is shown in Figure 7.22.

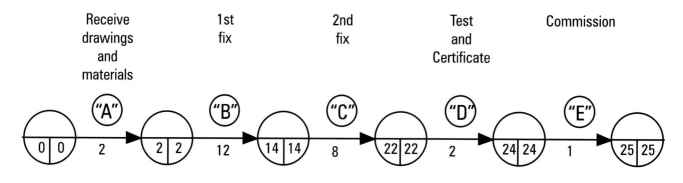

*Figure 7.22    Critical path network*

A typical network is, at first sight, somewhat complex but it is really no more involved than a regular bar chart, it's just not as user friendly. In a typical critical path network we show each task as an arrow with the base of the arrow as the start and the point as the finish.

A critical path network is able to show diverging and converging activities in a way not possible with a bar chart. The network provides a means of calculating the time to complete the project, based on the time taken for each task and their sequential relationships. The route through the network which takes the longest time to complete is the critical path. Should the tasks or activities along this path be disrupted or take longer to complete then the project will be delayed and will not finish on time. Any activity which is delayed for too long will become a critical path item as it will eventually affect the completion of the project.

The network provides a valuable management tool with which to monitor and control the project as it progresses. It is not however, in the most easily read format. This is usually overcome by the production of a bar chart programme which accurately reflects the network as a series of bars related to each activity and duration.

*Figure 7.23      Critical path network and bar chart programme*

# Administration

In addition to the programme and monitoring the progress of the work we must also use the site administration and records to control the project. We are involved in all types of paper work and some of these play an important role in the control of the work.

## Site records

Whilst we are working on site, it is inevitable that we will receive instructions that will affect our work or method of operation. For example we may be asked to change, omit or install additional materials, plant or equipment, change our programme of work or vary our working hours.

The paperwork recording these instructions plays an important part in the control of the work, each one having an effect on the progress and cost and possibly altering the original contract. There is generally a requirement within the contract to respond to an instruction notifying the client of any cost or programme implications within a prescribed period, often seven or fourteen days.

Any instruction which changes the original requirements agreed in the contract or specification must be issued by someone with the authority to do so. This will generally be the client's designated representative, for example the main contractor, the consultant or architect. It is usual for all such instructions to be issued through the same company so all variations will come from the same source, irrespective of who requires the change. Changes out are usually instructed by way of Variation Order and/or Architects Instruction.

| Variation Order | | |
|---|---|---|
| Issued by: George Edward Associates | | |
| Employer: Rachel Louise Hairdressing | | |
| Contractor: Arual Developments | | |
| Works situated at: Rachel Louise Hairdressing, Linking Road | | |
| Job Ref: C535 | | |
| Issue Date: 01-04-0- | | |
| Under the terms of the above contract, I/we issue the following instructions: | | |
| Instruction:<br>1.1 Add to contract No. C535, 2 further 13A sockets in the reception area. Quotation from S.Holly & Sons (Electrical Contractors) attached.<br><br><br>Signed: | Office use:<br>£ omit | Office use:<br>£ add |
| Amount of Contract Sum £ | | |
| Approx. Value of previous instructions £ | | |
| Approx. Value of this instruction £ | | |
| Approx. adjusted total £ | | |

*Figure 7.24      Variation order*

## Site diary

The site supervisor for the electrical contractor is responsible for keeping a site diary which is used to record all the happenings on site each working day. The actual format of the site diary will vary from company to company but Figure 7.25 shows a typical arrangement.

| CT Electrical Contractors | | | | No. | |
|---|---|---|---|---|---|
| SITE DIARY | | Contract Contract No. | | Date | |
| PROGRESS – (Work done, location and if in advance, on schedule, or behind | | | | | |
| DELAYS – (Overdue materials, other contractors, etc.) | | | | | |
| MEMOS – To Depts., Managers, etc. (brief details of contents) | | | | | |
| MISCELLANEOUS – 1: V.O. Necessary; 2:Overtime (reasons); 3: Surveying (info.request); 4: Visitors; 5: Other business | | | | | |
| PERSONNEL ON SITE: F/M C/H TECH AE ELECT APP JOINTERS LAB | | WEATHER: | | ACCIDENTS: Witness | |
| Date | | Signed | | | |
| DISTRIBUTION AND PRIORITY | | | | | |
| Priority | Title | Name | Priority | Title | Name |
| | Divisional manager engineer | | | Contacts manager | |

*Figure 7.25      Site diary*

The standard details recorded are usually address, job number and client, day and date. In addition the diary should record the name and designation of all the electrical contractors personnel on site, details of telephone and facsimile calls made and received and deliveries received and any goods returned.

The diary should also contain records of site meetings, instructions and drawings received and issued, information requested from the client and others and any incidents or occurrences on site together with any disruption to work or areas completed or handed over and the like. Some of these items may be recorded in separate documents, but the whole goes to make up the site diary.

Whilst this list is not exhaustive, it does show the type of information that needs to be recorded. This document can prove to be most useful in the event of any dispute as to when things happened, who said what and other contentious items. An experienced site manager will know instinctively what needs to be recorded because of the possible repercussions later. It is a most valuable and important record and should never be overlooked.

## Reports

In addition to the usual reports on the progress and activities on site which are part of the effective site operation, it is also necessary for incidents occurring on site to be reported. As the list of items required in the site diary does not leave a lot of space for recording events where there is much to record many companies have a standard report form for these occasions.

| CT | | Distribution | |
|---|---|---|---|
| | | Action | Information |
| Job No. | | Ref:    Date: | |
| To: | | | |
| From: | | | |
| Subject: | Site: Reason: Authority: | | |

**REPORT ON OCCURRENCE/ACTION/REQUEST**

| Numbered items | Action by |
|---|---|
| | |

*Figure 7.26      Typical report sheet*

**Remember**
The site diary(s) should remain secured on site at all times as they are the record of the site activities. They are the first line of reference for the site manager in the event of a query.

## Points to remember ◀ — — — — — — — — — —

All work undertaken is carried out under some form of contract, however basic.

There are distinct advantages to all parties in agreeing a formal written contract.

The contract agreement begins at the time the contractor is asked to provide a quotation or tender for work.

Contracts are subject to contract law and on larger projects can contain some onerous conditions on the parties involved.

There are a number of conditions which may affect the contract being made including lapse due to time, withdrawal, death of the contractor and rejection of an offer.

It is advisable to seek professional advice when entering into a contract.

The principal parties involved in the contract process are the client, architect, consultant, management contractor, main contractor and subcontractor.

The electrical contractor is usually a subcontractor.

Within the contract there is an important element of administration which must be undertaken. This involves the planning, programming and monitoring of all aspects related to the contract including work progress, labour, material, deliveries, changes to the original work, alterations and additions and other trades' activities.

It is important that the electrical contractor maintains accurate site records related to all the site activities.

The use of variation orders, site diaries and reports all go towards recording the activities on site and their effect upon the performance and progress of the contractor.

## Self-assessment short answer questions

1. List the conditions which must be met if a contract is to exist.

2.  List the principal reasons for a contract not being formed.

4.  State the key elements which need to be considered when planning activities on site.

3.  List the parties who may be involved in the contract process.

5.  Two principal progress monitoring charts are used on larger projects. Identify the two and briefly state the purpose of each.

# 8

# Cost Effective?

Before you start work on this chapter, complete the exercise below to ensure that you remember what you learned earlier.

All work carried out is under some form of contract no matter how basic.

All the parties involved in the work benefit from having a formal contract in which the details of the _____ and _____ are recorded.

The contractor makes an offer to the client in the form of a _____. The client then has to make an unqualified acceptance of the offer and the contract is _____.

There are a number of conditions which can prevent a contract being made or _____.

The parties involved in the contract process may include the client, architect, _____, management contractor, main contractor and subcontractor.

The planning and _____ of the progress on site forms an essential part of the contract_____ and is _____ if the project is to be completed on schedule.

Bar charts and _____ _____ networks are used to plan and monitor the progress of the project.

The electrical contractor needs to maintain _____ site records in order to identify the activities on site and any problems which may arise.

Typically a record of variation orders, a site diary and reports should be maintained by the supervisor on site.

## On completion of this chapter you should be able to:

◆ state the sources of contract cost data as related to materials, plant and labour
◆ identify opportunities for saving costs on material control, improving use of resources and applying new technology
◆ list the possible causes of delay to progress
◆ identify means of improving contract progress

# Part 1

# Information and instructions required

During the course of a particular project it is not uncommon to find that we do not have enough information about certain items. When this occurs we need to request details from our client and many companies have their own standard form or letter with which to request such information. It is essential that the request is clear and concise so that there is no doubt about the information being requested.

Similarly there may be occasions when we need to seek an instruction to vary our work. One example would be where we need to change the route of the electrical services to avoid a particular building obstruction, say a supporting beam, that was not identified on the drawings. Again many companies use a standard format for this and it is important that we state exactly what is required. If we are able to do so, dependent on the urgency and extent of the operation, we should notify the cost and programme implications of the changes when the request is submitted.

*Remember*
When the instruction is issued it must describe exactly what is required and this detail will be based upon our request.

It is important when asking for information or instruction to state the latest date by which the response must be received if we are to remain on programme. This should take account of the time required for ordering, delivery and any installation constraints. If a response is not received by that date, we need to notify our client of a delay to the programme. This will need to be monitored, and the client advised on a regular basis, if we are not to incur additional penalties for delay.

We must maintain a register of the requests and this should contain a minimum of

- a request reference number
- date the request was made
- date by which a response must be received
- subject of the request
- date the response was actually received
- whether the response was satisfactory
- further action or request made (with the appropriate reference number)

This information can then be discussed at regular site meetings and the flow of information easily monitored

**INFORMATION REQUEST REGISTER**

From: · · · · · · · · · · · ·
Contract No.· · · · · · · · · · · ·
Contract Name: · · · · · · · · · ·

| Drawing no. | Request details | Request no. | Issue date | Required date | Actual date | Satis-factory response | Further action request no. |
|---|---|---|---|---|---|---|---|
| | | | | | | | |
| | | | | | | | |
| | | | | | | | |
| | | | | | | | |
| | | | | | | | |
| | | | | | | | |
| | | | | | | | |

*Figure 8.1    Information request register*

# Conditions of contract

Having already considered some of the documentation that we shall need to complete during the course of the work we should take a little time to review what is involved in accepting a contract.

If we agree to carry out a small job of work for a domestic consumer there is not usually any written contract. Due to the relatively small value of the work and the nature of the installation it is often not considered necessary. However,

written contracts offer protection to both the customer and the contractor. Simple forms of domestic contract have been produced and proposed as part of the DETR's "Combating the Cowboy Builders-Working Group Report" published in 1999.

Annex E

**Checklist for Household Repairs and Maintenance Works (for small and/or emergency jobs not normally exceeding £500 in value)**

THE CUSTOMER (name, address and telephone number)

THE CONTRACTOR (name, address and telephone number)

**Work to be done** (list any special requirements, including materials)

**Timetable for the work** (insert dates)

Start:.............................    Finish:.........................

**The Price** (based on a firm quotation; including VAT, if it applies)

£........................

**Payment Terms**

For small works, this is normally on completion of the work.

Customer's Signature:.................................................

Contractor's Signature:................................................

**NB.** If customers need help or advice before undertaking work on their homes they should contact their local authority, Citizens Advice Bureau or other local advice agency before signing this agreement.

In the event of a dispute the terms and conditions of the contractor's named default contract will apply.

ANNEX F

**The Quality Mark Scheme: Household Repairs, Maintenance and Improvement Contract**

1    Between....................................................(the employer - the consumer or householder)
of............................................................
and...........................................................(the contractor)
of............................................................
..................................................(full address and telephone number)

2    The contractor will carry out and complete the following work for the fixed sum of £.............plus VAT (if this applies) at £......................(itemised as appropriate where varying rates apply) to a total of £......................... (If there is not enough space below, refer to separate specification, including any separate designs by architects or other designers).

3    The work will start on ...../...../...... and will be completed by ....../...../....... The contractor will not leave the site for more than 5 days in a row without a reasonable explanation, and will carry out the work using reasonable skill, care and progress. The contractor will tell the employer straight away of the cause and extent of any delay. If the contractor cannot meet the original completion date because of things outside their control, such as bad weather or sudden illness, they will agree extra time with the employer for carrying out the work.

4    The contractor will provide everything necessary for, and be responsible for, carrying out the work properly and efficiently, including labour, materials and equipment, unless the employer says otherwise in writing. All materials will be fit for their purpose, and will be new unless the employer has agreed otherwise in writing.

5.    The following person is responsible for getting any necessary planning permission and building regulations consent and must make all notifications, arrange inspections and pay any application fees in connection with the work:

the employer        the contractor

6    The employer will, where practical, make sure there are no obstructions on the site, such as blocked paths or driveways, and remove all furniture, fixtures and fittings that are necessary for the contractor to carry out the work. The employer will provide the necessary facilities (for example, power and water) as long as the contractor gives them plenty of notice.

7    The contractor will only carry out variations to the work (for example, extra or different work) if they have written instruction from the employer, including agreement to extra costs and time for completing the work.

8    The contractor will take full responsibility for the work, including any work carried out by his subcontractors. They will put right, at their own expense, any loss or damage caused either by himself or his subcontractors. The contractor will also insure against any loss or damage to the work or materials under a contractor's 'all-risks' policy. The contractor will give the employer appropriate evidence of insurance if they ask for it.

9    The contractor will meet legal insurance requirements for their employees; and provide suitable cover against injury to third parties or damage to third party property under public liability insurance, to a minimum of £2 million. The contractor will give the employer appropriate evidence of

*Figure 8.2    Draft contracts*

Normally there is some discussion between the contractor and the client as to how the job will be carried out, the finish that is required and the equipment and accessories that will be used. Any dispute between the client and the contractor is either settled by discussion or taken through the legal system. Indeed the values of work involved in such disputes can often be sufficiently low to be dealt with in the small claims court.

With larger projects there is usually a fixed date for completion and the client, or the client's representative, will usually set down conditions related to:

- the execution of the works, and the associated responsibilities and liabilities

- the issue, receipt and actioning of instructions

- loss or damage

- responsibility for plant and equipment

- completion of the work and extension to time

- failure to complete on time

- payment

- disputes and arbitration

- fluctuation in labour, materials and tax

This list is not exhaustive and covers only the major items, many others are included in the contract documents but we shall not be considering them at this time.

The contract places constraints upon what we can and cannot do. The determination of the implications and effect of these terms and conditions is a highly specialised job. Any disagreement or dispute will need to be dealt with under contract law.

As the electrical work is usually subcontracted from another party, it is common for the work to be under one of a number of standard forms of subcontract used for these arrangements, typically Domestic Sub-Contract DOM/1 and the JCT 80. (Domestic in these terms does not mean "domestic" as applied to dwellings) Other forms of contract are available and used for works contracts, produced by parties such as Government Bodies and The Association of Consultant Architects.

*Figure 8.3    Typical subcontract form*

We have already considered the requirements of the subcontract and the implications that this will have on our operations. When we accept the contract and its conditions we will find that one of them refers to the documents issued for the purpose of providing the tender. It is important that we are aware of exactly what was issued for tender and any additional documents used in the tender process. As a general rule these will include a minimum of

Tender Drawings

Specification

Tender Programme

Contract Conditions (where these are not a standard form of contract)

Any correspondence or further details issued during the tender period

Once the contract is awarded a set of contract documents is normally issued, including a copy of the working drawings, any revised specification and the contract conditions. The first task will be to establish if there are any differences between the tender documents and the contract issue of documents and drawings. Any differences are likely to have some implication in respect of material or time and therefore will affect the cost of the work. Any such changes must be recorded and notified to our client on the standard forms, including the implications to cost and programme.

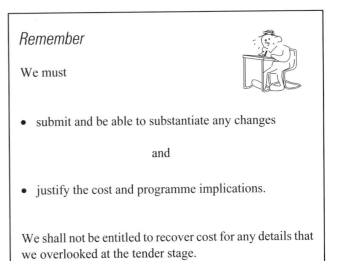

Remember

We must

- submit and be able to substantiate any changes

and

- justify the cost and programme implications.

We shall not be entitled to recover cost for any details that we overlooked at the tender stage.

Once we have accepted the contract and advised our client of any changes to the original tender we must ensure that the cost involved in completing the work is controlled. The better the cost control the better the chances of making the necessary level of profit, thus ensuring the job is successful, improving the company's success and securing our own jobs.

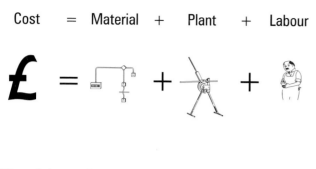

$$Cost = Material + Plant + Labour$$

Figure 8.4    Cost equation

Figure 8.5    Price advantage

The main cost factors that we need to consider are

- material
- plant and equipment
- labour

so in this chapter we shall consider each item and the means available to us to maintain control over the costs involved.

# Material cost control

Within the contract and specification there are usually references to the type and manufacturer of the principal items of material, such as lights and accessories. Unless we have specifically agreed a change to these items, we are obliged to provide what was stated in the tender. Some major items of plant and equipment may also be specified as to the manufacturer. In which case, unless we have specifically agreed a change with the client, we are obliged to provide what is specified.

The requirements for other items of material and equipment may be a description and reference to a particular standard. For example the type of cable to be used may be specified as "BASEC Approved LSF cables". This type of requirement allows the contractor to select any manufacturer who provides a product to meet that specification. Ancillary items, such as screws, wall plugs and the like, are generally not specified, unless there is some particular requirement. These items are simply expected to be suitable for the intended job and the environment in which they are installed.

The control of material cost falls broadly into three categories;

## Purchasing advantage

This is where the contractor negotiates the best possible purchase price for the material required. This may be negotiated on the basis of quantities, delivery and supplier's discounts, although it may prove difficult to negotiate cost savings on equipment where the supplier has been specified.

## Material control

The use of the material on site must to be controlled in order to minimise wastage and theft. Many companies use a booking out system for material, particularly the expendable items of material and accessories, to enable them to control use and wastage. These allocations per area can be cross referenced to the record drawings and site measures to help control material cost. Storage and handling are also important aspects in reducing material wastage due to damage and theft. The principal storage should provide secure and suitable environmental conditions to ensure that the materials and equipment are not damaged or stolen. Once the material makes its way to the work-site there should be secure storage available, such as site boxes, to allow the materials to be stored safely if the site area is to be left unattended. Operatives should be reminded of the need to secure materials and minimise their wastage during the installation activities.

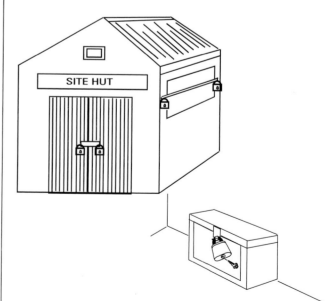

Figure 8.6    Suitable secure storage must be available

## New technology and change

Where new technology or alternatives become available it is possible to recommend a change to the material or equipment to the client. However if the change is to be acceptable, it is likely that the client will expect to see some benefit. This generally means that some of the cost saving will have to be passed on to the client.

# Part 2

## Plant and equipment costs

The principles of negotiation with regard to the cost of the plant used for the construction of the installation are similar to those used for materials. Contractors must consider whether it is economical to hire plant or more cost effective to purchase items of plant. In general terms it is more economic to hire plant which is only going to be used for short periods of time or has a very high initial cost. A mobile scaffold used throughout the whole of the construction of the electrical installation may be more cost effective to purchase, whilst a fixed scaffold to the exterior of a building may be more economical to hire.

The basic formula is to establish the period of time for which the item is required, establish the most competitive cost for hire and compare it with the cost of purchase. The company may make a decision that, as the plant will be used frequently on later projects, purchase may be the better long term option. In respect to our individual project costs, apply the formula and use the cheapest option.

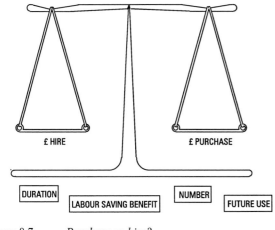

*Figure 8.7      Purchase or hire?*

Further control of the cost of the plant is achieved by ensuring that the plant is only on site for the minimum period necessary to complete the work. Careful planning and distribution of labour are essential if we are to keep the plant costs to a minimum.

The plant used in the construction of the installation is linked directly to the labour cost. Whilst the initial cost of some plant may seem high, the saving in man hours far outweighs the additional plant cost. For example, the old fashioned wooden conduit setting block is much cheaper than a conduit bending machine. However the time saved in the production of a number of conduit bends of identical size, shape and radius provides a considerable saving in labour cost. This results in the bending machine being the cheaper option. New technology may also play a part in the cost of plant and new equipment and may offer some considerable cost savings which should not be ignored. For example the use of laser setting out equipment for long, straight runs of lighting.

## Labour costs

The time it takes to complete a job of work can be measured and predicted with reasonable accuracy. It is the "interruption and delays" to the work activity which cause the time to overrun. The setting up of the work stations and the packing away at the end of each day, time taken for prescribed work breaks and like activities, all affect the length of the productive working day. Added to this are such factors as obstruction, congested or restricted work areas, difficult access and permitted working time due to the activities of others. All of which conspire to add labour cost to the work. Contractors need to ensure that delays and obstructions, which are within their control, are kept to a minimum.

One way of helping to reduce these short term delay costs is to have activities, or areas of the site which are not critical. These locations are where operatives, who are prevented from working for one reason or another, can achieve some productive work at short notice. Such areas are often referred to as "hospital areas" as they are where the "casualties of construction delays" are sent to make best use of their time.

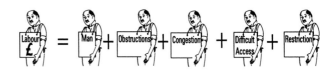

*Figure 8.8      Labour costs*

# What factors affect cost?

The cost of completing the work may be affected by a number of elements. In the main the major cost is caused by delay to progress, coupled with the cost involved in recovery. As a result we need to consider the main causes of delay and the implications of each. Major items which we shall consider are;

- contract anomalies
- shortage of resources
- site constraints
- environmental conditions
- third party actions

# Contract anomalies

We have already stressed the importance of ensuring that the contract covers all the agreed aspects of the work. Failure to identify anomalies with the client at an early stage can result in delays, disruption and additional cost later. Prevention is much better than cure and it is in our best interest to bring to the client's attention any omissions, or items which appear to be conflicting, as soon as they are discovered.

For example, if you are aware that there are a number of air handling units being installed, but the electrical installation associated with them is not in your contract, then it is important to bring this to the attention of the client. If no allowance has been made within the electrical contract for this work, because it has been let to the company installing the air handling units, then the installation aspects may have been covered. However, if no allowance has been included in the distribution board arrangements for supplies to this "other work" then changes will need to be made. This will usually result in some delay, unless the situation is identified early enough to ensure the correct equipment is installed first time.

*Figure 8.9*      *Contract anomalies*

## Shortage of resources

This applies to both labour and material. We need to carefully plan our work to ensure that the correct level of labour and sufficient material is on site at all times. The control of the labour and their associated material is vital to the cost effectiveness of the job. Too little labour and the progress will begin to fall behind schedule. The cost of additional labour, or

the overtime worked by the existing labour, will put the job over budget. Likewise too much labour will result in increased cost without an associated increase in productivity. Too many operatives and not enough work sites or overcrowding will reduce efficiency. Breaks will take longer due to the increased numbers using the facilities and, whilst staggering the breaks may help, the time taken inevitably increases. So we need to monitor progress and maintain the correct level of labour on the site at all times.

The need for a clear programme for the delivery and availability of the necessary materials and plant to allow the operatives to work effectively and efficiently also becomes apparent. If we have the correct level of labour available then it follows that there must be sufficient work areas, plant and equipment and material for them to operate efficiently. Preplanning of the level of all resources is therefore one of the key issues in the cost effectiveness of any job.

---

*Remember*
Any shortage or delay in the provision of the plant, material, equipment or operatives to adequately resource the electrical contractor's work will result in a delay to the programme. This will inevitably result in additional cost as the contractor attempts to bring the work back to programme.

---

The time penalties on large contracts may be so high as to make an exceptionally high level of additional resource, to meet the programme deadline, the least costly option. Every effort must be made to prevent delays to the progress of the job being caused by the electrical contractor's failure to adequately resource the work. Similarly the resource must be maintained within the budget for the contract. Careful planning and control are the keys to a successful project.

*Figure 8.10*      *Resources shortage*

## Site constraints

In addition to the physical constraints of the site, such as access corridors, doorways, obstructions and restricted working areas, there are some additional factors which must be considered if we are to control the progress and cost of the installation effectively.

The location of a site and the operating requirements of both the site and the surroundings may give rise to delays in the progress of the work. Some of the restrictions on operation are often recorded in the original tender documents as they will have a significant effect on the time required to carry out the work. The type of restrictions usually mentioned are noise limitation, access, delivery and operating hours.

Certain types of site have restrictions on the hours during which noisy activities such as percussion drilling, chasing and demolition can be undertaken, due to the working requirements of adjacent tenants. This is particularly common when refurbishing a single floor or floors in an existing office block where the other levels are still in use.

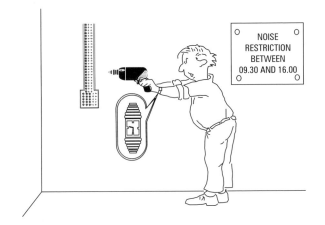

*Figure 8.11      Site constraints – working hours*

Inner city projects often have limited access for both plant, equipment and operatives where good planning and liaison can help to reduce the likelihood of delays occurring. This may mean that the operatives and subcontractors will not be able to park vehicles on site and may need to carry equipment and tools some distance. Deliveries may need to be strictly scheduled and a missed time slot would then result in no delivery being possible until the next vacant time period. Some sites may have to restrict deliveries to outside of working hours only.

Such constraints may not only incur additional cost to meet the delivery times, but failure to make a delivery will cause delay to the progress of the site and the productivity of the labour. It may also be necessary to arrange for road closure notices and the associated costs for the delivery and unloading of large items of plant and equipment.

Of course the problems are not restricted to just the delivery of materials and equipment. Rubbish, packaging and redundant plant and equipment will need to be removed from the site. It may not be possible to arrange for a skip to be located on or adjacent to the site. Waste may need to be placed into sacks and stored, with regular pick-up times within the site access schedule.

Any such restrictions have a real potential to cause delays to the progress of the work and incur additional cost to those associated with the actual task. Any missed opportunity in the access schedule can seriously disrupt and delay the site activities attracting considerable additional cost as a result. Careful planning and scheduling of these activities are important factors in the control of the project progress and cost.

## Environmental conditions

Many of the activities undertaken on site, particularly during the early stages of construction of a new site, are subject to the elements. Construction is often delayed due to the weather, particularly rain and frost, and such delays will affect the whole construction programme, and all the activities which follow.

The environment can also play a part in the electrical installation programme by directly influencing the installation activities. For example the installation of PVC cables, conduit and trunking is affected by the extremes of heat and cold. In buildings with no heating, insulation and often no windows installed the temperature may fall too low to allow the installation of such materials. Similarly the rise of temperature within a glazed area may rise too high in mid summer for the installation of such materials to take place.

*Figure 8.12      Site constraints – environmental causes*

Not only may the environmental conditions affect the installation of material, they may also have an effect on the operatives carrying out the installation. Even when an operative is suitably equipped to carry out a task in particularly inhospitable environments, the protective equipment often causes restriction in movement and agility, resulting in the task taking longer to complete.

## Third party actions

Working on an isolated site, where we are the only contractors carrying out work, would seem to guarantee that our costs and programme are not unduly affected by the actions of others. However if the highway maintenance programme closes the access route for road resurfacing the time to and from the job could be considerably increased, either due to traffic delay or diversions.

Where our activities are in locations where there are other trades and the adjacent or surrounding areas are not under the control of the site managers, the potential for delay is greatly increased. The office next door deciding to have work carried out at the same time, the supply industries deciding to install new main services in the access roads, and the like are quite common occurrences.

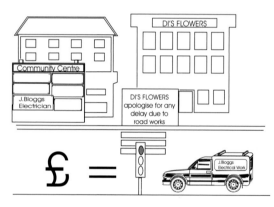

*Figure 8.13      Third party actions*

The schedules for activities on the site should be controlled to prevent trades causing too much disruption to each other. Unfortunately, if for some reason one stage of the project is delayed then, to catch up, the various trades working times are usually reduced and overlapped. This often results in the trades sharing areas, and some activities such as floor laying, ceiling erection and painting can seriously hinder the progress of other trades.

With careful planning and management it is possible to establish if any other activity is proposed in the locality which may affect our project. There will always be the risk of some other activity being undertaken unannounced which will cause some disruption to our progress. When this does occur, we must take sensible cost-effective measures to overcome the problem. This may involve the whole project team and a

workable solution must be established as soon as possible to minimise the effect on the programme and the cost of the project.

# Improving progress

The requirement for contracts to be completed on time and on budget is fundamental to the success of everyone involved. For the electrical contractor the ability to remain in business may be threatened if the works undertaken are not profitable. It is therefore vital for the survival of the company to ensure that the contract is a success. Ensuring the best progress throughout the job involves frequent review and prompt action to redress any shortfalls in the programme and progress. Improving the progress of the work can be achieved by considering the following principal requirements

## Planning

The planning of the work and resources is vital for the success of the project and should be carried out from tender stage right through to the completion and handover to the client. Every aspect of the work relies on a number of requirements;

- the preparation of the work area
- the provision of material and equipment
- provision of labour
- access to the work site
- time to complete the task

Careful planning of each aspect is crucial to the success of the project. Should any of these requirements, which are beyond our control, threaten to affect our work then the client should be advised, with details of the cost and progress implications, without delay. The activities which are normally outside our control are those not contained within our contract, such as the preparation of the work area and access to the work site. The supply of our material, plant, equipment and labour are usually our responsibility and failing in these requirements must be dealt with promptly if we are not to become involved in additional costs.

*Figure 8.14      Improvements*

## Resourcing

Monitoring the progress of the work and ensuring the level of resource required to remain on programme is essential. Planning the work results in an anticipated demand for labour and resources, which should be monitored as the work progresses. If it becomes apparent that the levels of resource need to be changed, early programming for these changes will minimise any problems. It may be that the work is progressing faster than was estimated and so deliveries will need to be brought forward and labour levels adjusted to suit. If progress has been delayed then it may be that deliveries need to be put back, and less labour required, than originally expected. It is also quite probable that at a later stage of the work additional resources will be required and aspects of the work will need to be reprogrammed.

A register of requests for information should be maintained identifying:

- the date the request was made
- the nature of the request
- the date for reply
- the date reply received
- the suitability of the information.

Draft contracts have been produced, through the DETR, for use in connection with domestic installations. Larger sub contract works are usually carried out under standard forms of contract such as DOM/1 and JCT80.

It is important to establish any change between the information provided as part of the contract and that provided for tender. The main cost factors which need to be controlled are material, plant and equipment and labour.

Material cost can be controlled by purchasing advantage, control of material and the use of new technology where possible.

Plant and equipment cost needs to be considered in terms of purchase or hire.

Labour cost needs to be controlled by careful planning and distribution of labour, plant and material.

Factors which affect cost are:

- contract anomalies
- shortage of resources
- site constraints
- environmental conditions.

Careful programming and planning help to control and improve progress and cost.

## Self-assessment short answer questions

1. List the minimum detail that should be recorded in a register of requests to the client.

2. List the three main factors that need to be considered in the control of material costs.

3. List the considerations which affect the cost of plant and equipment.

4. List the factors which affect the contract cost.

5. List the main factors which can help reduce costs associated with the contract.

# 9

# Maintaining Good Relationships

Before you start work on this chapter, complete the exercise below to ensure that you remember what you learned earlier.

An information _____ register should be maintained identifying the date the request was made, _____ of the request, date for_____, date reply received.

Draft _____ have been produced for use in connection with _____ installations.

Larger sub contract works are usually carried out under _____ _____ of contract.

It is important to establish any change between the_____ provided at tender and that provided with the _____.

The main cost factors which need to be controlled are _____, plant and equipment and _____. Material cost can be controlled by_____ advantage, control of _____ and the use of new _____ where possible.

Plant and equipment cost needs to be considered in terms of _____ or hire.

Labour costs need to be controlled through careful _____ and _____ of labour, _____ and material.

Factors which affect cost are contract _____, shortage of _____, site constraints and _____ conditions.

Careful _____ and planning help to control and _____ progress and cost.

### On completion of this chapter you should be able to:

◆ describe the information which can be provided to customers
◆ recognise the means of checking the awareness of customers
◆ explain the importance for the organisation of good relationships with customers
◆ explain the importance of effective relationships with colleagues and other parties

## Part 1

## Maintaining good relationships

In the earlier books in this series we discussed the need for good working relationships with customers, colleagues, other trades and the public at large. At this point it would be worthwhile to remind ourselves of the basic requirements.

A business cannot succeed without its customers and the success or failure of a company can be determined by the way in which it is represented by the staff. The aim is to provide a caring, polite and professional image and deliver a first class service

## First impressions

First impressions of a company count as they are often the ones by which the company is judged and upon which future custom depends. The impression given will, of course, vary dependant upon the circumstances but the aim should be to provide a positive impression at all times.

*Figure 9.1*     *First impressions are important!*

The basic considerations are;

Are the staff

- clean and tidy (conditions permitting)?
- polite?
- helpful and knowledgeable?
- on time for appointments?

And do they

- treat customers with proper care and consideration?
- take the right tools, materials and equipment to the job?

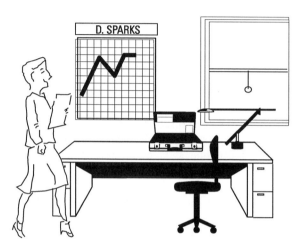

Figure 9.2    A smart and welcoming office

Is the

- company office clean, tidy and welcoming?
- workshop tidy and well kept?
- company van and plant clean and in good condition?

Figure 9.3    A smart van

Are

- accurate records kept and readily available?
- messages and information relayed?
- enquiries dealt with efficiently, effectively, correctly and politely?

**THIS**

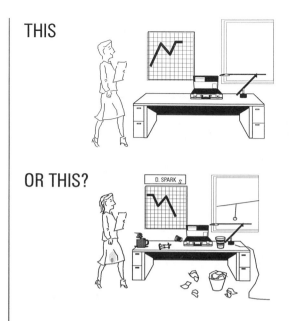

**OR THIS?**

Figure 9.4    An efficient, well kept and tidy office will give customers' confidence in placing an order

Care in these areas of a company's operations is vital if customers are to be confident in placing the work and able to recommend the company to others.

Once the work has been obtained, it is equally important to maintain a good working relationship with the client. One of the major stumbling blocks in maintaining good customer relations is the lack of information provided to the client. It is therefore essential to establish the client's understanding of

- the extent of the work
- technical implications
- programme implications
- any actions required

from the information provided.

As we have said before, a diplomatic approach in the early stages is important to determine the level of understanding without causing offence. Once the initial level of understanding is established, we must take care to keep the client informed in terms which are appropriate to the level of technical understanding. So the terminology used will be different when communicating with an electrical consultant compared with a layman.

Figure 9.5    Communicating with the client is important

Having provided customers with the appropriate information, it is also important to make sure that they understand what we have provided. This also needs to be carried out in a considerate and tactful way so as not to cause offence. Allowing the customer to ask any questions regarding the information may not be sufficient, particularly when dealing with customers who are non-technical. It is often a good idea in those cases to offer some brief verbal explanation outlining what has been provided and, where the customer appears to be uncertain, offer some further detail.

Figure 9.6        Electrician providing an explanation to the customer

Remember
It is important to make sure that all your dealings with the customer are carried out in a polite and courteous manner.

Do not talk down to the customer.

Do not provide explanations in very technical language unless the customer is familiar with the electrical installation industry.

The ability of an individual and company to maintain good relationships with their customers is vital to the success of the company. If the customer is pleased with the service provided, is kept informed throughout the work and finishes up with the standard of job they expected, then the effects may be twofold.

The customer may provide repeat business, so any further work which is required would be passed to the company for tender. The bid may even be considered favourably in the light of previous experience. This can be particularly important where the customer is a builder, developer, consultant or management contractor, and where the work carried out was part of an overall development or refurbishment programme.

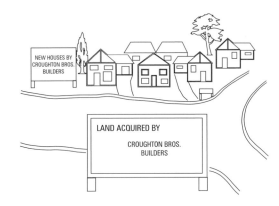

Figure 9.7        Repeat business may be the result if a customer is satisfied with the work

The customer may also provide a recommendation for the company to others who are considering having electrical work undertaken. This can be particularly significant where the market is for smaller works such as domestic or minor and maintenance work. A recommendation from a personal acquaintance or colleague is often the most effective advertisement for a company and such recommendations are only given by satisfied customers. Satisfied customers are only obtained by the provision of good quality information, a high standard of service and a first class job.

Figure 9.8        Satisfied customers will recommend a company

Remember
A good reputation is hard won and easily lost. It is vital to maintain good customer relationships from the time of first contact or enquiry to the final completion and handover of the work.

# Part 2

## What information should we provide?

The customer has a right to information regarding the work they have requested, and this extends beyond just the cost and the time taken to complete the work. We have already considered the need to inform the client of changes to cost or programme and so we shall not be covering these again here. However, dependant on the type of work and the information provided at the time of tender, the contractor does have an obligation to provide the customer with certain information. Similarly the customer has a right to be given certain information.

## How do we provide this information?

Dependant upon the type of job and information provided by the client the information may be provided by means of;

### Specification

Where the client does not provide a specification then the contractor should provide one to the client detailing what the job will comprise. Typically a domestic consumer requesting a rewire will have an idea of where lights and sockets are required but that is generally the extent of the client's input. The contractor should provide a specification of the work to be carried out which includes;

- the number of points provided in each room, lights, sockets, telephone, TV and the like
- the make or standard of accessories to be used
- the extent of the work to be undertaken
- any additional items, such as outside lights
- any exclusions such as rewiring the central heating or where light fittings are provided by the customer
- cost and programme

The above list is not exhaustive but it does provide an indication of the minimum information that should be provided to the customer in the absence of a detailed client's specification. This information is usually provided when the quotation for the work is submitted to the client.

*Figure 9.9    Information should be provided to the customer*

### Manufacturers' data for use and maintenance

Where items of equipment are installed, such as electric showers, boiler control systems, heating appliances, security and alarm equipment and the like, then the client should be provided with the manufacturer's information for the operation and maintenance of the equipment.

On larger contracts this information is included in the Operation and Maintenance Manuals and the provision of these is generally a requirement of the contract. The manuals are required from all trades whose installation or equipment needs to be operated or maintained.

The information should include details for the user on the everyday operation of the equipment. There should be information regarding any user maintenance required to keep the equipment operating effectively and efficiently.

There should also be information regarding the periodic maintenance of the equipment. This should include the period between maintenance activities and the extent of the maintenance required. In much the same way as a car service record sheet is produced, detail the items which require attention or replacement at each visit.

Circuit, block and wiring diagrams should also be provided in order that any maintenance or adjustment can be carried out safely and effectively. The information provided is essential for fault finding and, as each system may be different, it is not possible for technicians to carry every conceivable permutation. Each site therefore keeps its own O & M manuals which are made available to the residential maintenance staff and any specialist engineer called to the building.

*Figure 9.10    Operation and maintenance manuals should be handed to the client*

We must not forget that good working relationships should be maintained with all those involved in the process, not just the client.

# Colleagues and others

Good relationships with customers are essential to the success of the company, but that is not the only area where relationships are important. In order for the company to operate effectively and efficiently the employees must be able to work with each other as a team. It is also important that good relationships are maintained between companies involved in working on the same project. Animosity and ill feeling between colleagues and companies can seriously affect the progress of the work and the success of the project.

Many of the points highlighted for good relationships with the customer also apply to maintaining good relationships amongst the employees. Keeping the colleagues and other parties informed of what is going on helps to

- promote an understanding of the overall project, the progress and activities

- minimise jealousy and ill feeling about favouritism and the like

- improve cooperation

- encourage a good working environment

If the working environment and atmosphere are good then productivity and cooperation improve. The reverse is also true and the importance of maintaining a good working relationship, with all those involved in the project, cannot be stressed enough.

Figure 9.11    *It is important to maintain a good working environment*

Relationships with other contractors involved in the project should be maintained throughout the work. It is important that the trades work well with one another if the project is to be successful for all those involved. As we discussed earlier poor relationships with other trades can be particularly disruptive to the progress and cost effectiveness of the job. It is always beneficial to have good communications with the other trades, simply talking to them and keeping them informed of activities helps to develop a good rapport.

When changes occur, or particular actions which are likely to cause disruption. It may be beneficial to forewarn the other trades, which are affected, verbally in advance of the written details. This gives those concerned the maximum time to consider their alternatives and actions during periods of disruption. Fostering such relationships will also serve to help in our own progress as similar courtesy may be offered when actions by others will affect our work. This will allow us to be prepared well in advance, minimising our disruption.

Figure 9.12    *Mechanical and electrical contractors resolving a problem.*

*Remember*
Good working relationships with all those involved are vital for the success of the project for all concerned. Poor relationships result in inefficient working. Additional cost through delays and inefficiency and can often result in the client being dissatisfied with the progress and end result.

**111**

## Points to remember ◀------------

First impressions of a company are important as they are the ones by which the company is judged and future custom depends.

Staff should be tidy, polite, helpful and on time.

Customers should be treated with care and consideration.

The company premises should be clean and tidy and the staff welcoming.

Accurate records should be kept, with messages relayed and enquiries dealt with effectively and politely.

Good communication with the client is important to the success of the project.

Customer communication should be carried out tactfully and diplomatically.

The customer has a right to be provided with information before, during and on completion of the work.

Make sure the customer understands by using appropriate terminology to their level of understanding, without belittling the individual.

The success of customer relations has a direct relationship to the success of the company and continuation of work.

A good reputation is hard won and easily lost.

Aim to provide good quality information, a high standard of service and a first class job. Information is provided to the customer initially through the use of a specification and upon completion by use of manufacturers' information and forms of certification.

Keep the customer informed throughout the work.

Good working relationships with colleagues and other contractors are an important factor in the success of a project for all concerned.

## Self-assessment short answer questions

1. List six important factors related to the image the company should create.

2. List the ways in which the client is provided with formal information.

3. List the typical information that should be provided in a specification to a domestic client for a rewire of their premises.

4. Customer relations are important to the company. List the other relationships which are equally important for the success of the project.

5. List the probable effects of a poor relationship with those on site on the electrical contact.

# End Test

**Circle the correct answers in the grid at the end of the multi-choice questions.**

1. Which of the following does **NOT** form part of the contract team
   (a) client
   (b) supplier
   (c) main contractor
   (d) electrical contractor

2. A disregard for the safety of yourself or others can result in prosecution by the
   (a) client
   (b) police
   (c) health and safety executive
   (d) main contractor

3. The information related to the actual work done on a small job carried out on a time and material basis, would be recorded on a
   (a) time sheet
   (b) job sheet
   (c) part order
   (d) finalisation order

4. A common, easily interpreted, method of planning and monitoring progress on site is by the use of a
   (a) pie chart
   (b) bar chart
   (c) critical path network
   (d) progress register

5. In order to carry out a site survey contractors should ensure that they have the appropriate
   (1) tools
   (2) access
   (3) plant
   (4) equipment
   (5) insurance
   (a) only items 1, 2, 4 and 5
   (b) only items 1, 2, 3 and 4 are required
   (c) only items 1, 2, 3 and 5 are required
   (d) all of the above are required

6. Where large or heavy plant and equipment are to be installed, it is usually necessary to have a .............. survey carried out to confirm the final location is suitable.
   (a) site access
   (b) visual
   (c) dimensional
   (d) structural

7. Working adjacent to trenches and earthworks is subject to the same requirements as working
   (a) under a permit to work
   (b) in a confined space
   (c) at heights
   (d) underground

8. Which of the following is **NOT** a consideration when determining the types of installation materials to be used for an electrical installation.
   (a) type of building construction
   (b) environmental conditions
   (c) site access
   (d) purpose of the building

9. An item of equipment to be installed in a wash down area with high pressure hoses in use would need an IP rating of a least
   (a) IP X6
   (b) IP X5
   (c) IP X4
   (d) IP X3

10. The IK code is used to identify the degree of protection an item of equipment affords to
    (a) immersion in water
    (b) exposure to heat
    (c) ingress of dust particles
    (d) mechanical impact

11. Temporary electrical installations
    (a) do not need to be installed to the same standard as permanent installations
    (b) are subject to the same requirements as permanent installations
    (c) are subject to additional requirements
    (d) are not under the requirements of the statutory regulations

12. A typical example of the application of a ductile material would be
    (a) cast iron
    (b) copper
    (c) PVC
    (d) glass

13. Which of the following is a ferrous material
    (a) copper
    (b) aluminium
    (c) mild steel
    (d) lead

14. A suitable fixing for use when fixing a distribution board on concrete blocks would be
    (a) wall plugs and screws
    (b) toggle bolts
    (c) butterfly fixings
    (d) masonry nails

15. The most appropriate form of documentation that should be issued following the completion of an alteration to a single circuit, which does not involve the installation of a protective device is the
    (a) Electrical installation certificate
    (b) Periodic inspection report
    (c) Schedule of results
    (d) Minor electrical installation works certificate

16. The most appropriate form of documentation that should be issued following the completion of an electrical installation which involves the installation of a number of new circuits is the
    (a) Electrical installation certificate
    (b) Periodic inspection report
    (c) Schedule of results
    (d) Minor electrical installation works certificate

17. The most appropriate form of documentation that should be issued following the inspection of an electrical installation which has previously been energised and put into service is the
    (a) Electrical installation certificate
    (b) Periodic inspection report
    (c) Schedule of results
    (d) Minor electrical installation works certificate

18. To ensure an electrical installation is safe to be put into service it should be
    (a) fully inspected, tested, verified and certified
    (b) visually inspected
    (c) tested
    (d) certified

19. The certification issued following the completion of electrical installation work
    (1) details the extent of the work for which the electrical contractor is responsible and
    (2) should only be provided at the client's request.
    (a) statement 1 is correct and statement 2 is incorrect
    (b) both statements 1 and 2 are correct
    (c) statement 1 is incorrect and statement 2 is correct
    (d) both statements 1 and 2 are incorrect

20. When reporting on an electrical installation which has been put into service there is a requirement to record
    (1) observations on aspects of the electrical installation which do not meet the current standards and
    (2) what remedial action is required.
    (a) statement 1 is correct and statement 2 is incorrect
    (b) both statements 1 and 2 are correct
    (c) statement 1 is incorrect and statement 2 is correct
    (d) both statements 1 and 2 are incorrect

21. In order for a formal contract to exist there must be
    (1) a formal offer made and
    (2) an unqualified acceptance of that offer
    (a) statement 1 is correct and statement 2 is incorrect
    (b) both statements 1 and 2 are correct
    (c) statement 1 is incorrect and statement 2 is correct
    (d) both statements 1 and 2 are incorrect

22. Failure by either party to fulfil the terms of a formal agreement will normally result in
    (a) legal action under civil law
    (b) legal action under contract law
    (c) legal action under criminal law
    (d) legal action under common law

23. The member of the contract team, engaged by the client to advise on the technical requirements is the
    (a) architect
    (b) consultant
    (c) main contractor
    (d) management contractor

24. Successful administration of any contract requires
    (1) careful planning and programming and
    (2) close monitoring
    (a) statement 1 is correct and statement 2 is incorrect
    (b) both statements 1 and 2 are correct
    (c) statement 1 is incorrect and statement 2 is correct
    (d) both statements 1 and 2 are incorrect

25. To ensure the completion of larger and more complex projects on time it is necessary to determine and monitor the
    (a) delivery schedules
    (b) the bar chart
    (c) the programme
    (d) the critical path network

26. Changes to the scheduled work, material or equipment that are imposed during the course of a project will
    (1) require official confirmation from the client in the form of a job sheet and
    (2) will have an effect on the cost of the project.
    (a) statement 1 is correct and statement 2 is incorrect
    (b) both statements 1 and 2 are correct
    (c) statement 1 is incorrect and statement 2 is correct
    (d) both statements 1 and 2 are incorrect

27. The electrical contractor's site supervisor is required to keep a record of the activities on site in the form of
    (a) reports
    (b) variation orders
    (c) time sheets
    (d) site diaries

28. During the course of the project there may be a need to request information from the contract team or client. The normal procedure for this is to issue a
   (a) request for information
   (b) a variation order
   (c) an additional works sheet
   (d) a site instruction

29. The basic conditions of contract are produced as
   (1) Standard forms of contract for sub-contracts and
   (2) Draft documents for domestic installations in dwellings
   (a) statement 1 is correct and statement 2 is incorrect
   (b) both statements 1 and 2 are correct
   (c) statement 1 is incorrect and statement 2 is correct
   (d) both statements 1 and 2 are incorrect

30. The three main cost factors that need to be considered in relation to any project are
   (a) labour, plant and material
   (b) plant, material and storage
   (c) labour, site facilities and storage
   (d) labour, plant and equipment

31. The costs incurred in connection with the labour on the contract need to be kept to a minimum. This may be achieved in some part by
   (1) providing a labour force in excess of that required to minimise delay and
   (2) recruiting local labour
   (a) statement 1 is correct and statement 2 is incorrect
   (b) both statements 1 and 2 are correct
   (c) statement 1 is incorrect and statement 2 is correct
   (d) both statements 1 and 2 are incorrect

32. Keeping material cost to a minimum may be assisted by using
   (1) purchasing advantage and
   (2) secure and suitable site storage.
   (a) statement 1 is correct and statement 2 is incorrect
   (b) both statements 1 and 2 are correct
   (c) statement 1 is incorrect and statement 2 is correct
   (d) both statements 1 and 2 are incorrect

33. Two factors which affect plant costs are
   (1) the relative cost of hiring or purchasing plant and
   (2) the saving achieved in terms of man-hours
   (a) statement 1 is correct and statement 2 is incorrect
   (b) both statements 1 and 2 are correct
   (c) statement 1 is incorrect and statement 2 is correct
   (d) both statements 1 and 2 are incorrect

34. Shortage of resources on site may result in
   (1) delays to the programme and
   (2) increase in costs to the electrical contractor
   (a) statement 1 is correct and statement 2 is incorrect
   (b) both statements 1 and 2 are correct
   (c) statement 1 is incorrect and statement 2 is correct
   (d) both statements 1 and 2 are incorrect

35. Where adverse conditions are involved in the course of the electrical installation work the provision of suitable protective clothing and equipment
   (1) will allow the work to be completed without the need for additional time and
   (2) is a requirement of health and safety law.
   (a) statement 1 is correct and statement 2 is incorrect
   (b) both statements 1 and 2 are correct
   (c) statement 1 is incorrect and statement 2 is correct
   (d) both statements 1 and 2 are incorrect

36. When carrying out electrical installation work on a large project it is essential for the success of the project that
   (a) all operatives wear corporate clothing
   (b) lunch periods are staggered
   (c) good communications exist between contractors
   (d) adequate parking facilities are provided

37. Which of the following is not essential to good customer relations
   (a) Staff are polite and courteous
   (b) Company offices are luxuriously appointed
   (c) Company offices and workshops are clean and tidy
   (d) Enquiries are dealt with promptly and efficiently

38. Good communication with customers at all stages is important to the success of the project and in securing possible future work. This communication should be
   (a) always in writing
   (b) always diplomatic and considerate
   (c) always given verbally
   (d) always providing full technical detail

39. The level of detail provided to the client on completion of the work should be sufficient to allow
   (a) operation and maintenance of the installation
   (b) the client to maintain the building without professional assistance
   (c) the running costs to be monitored
   (d) a full service history to be maintained

40. Information may be provided to the client by means of a specification when
   (a) the client stipulates the work is to be on a time and material basis
   (b) the professional team provides bill of quantities tender documents
   (c) the client is not to be the eventual occupier of the premises
   (d) the client does not provide details for the installation beyond the position of accessories.

## Answer grid

| | | | | | | | | | |
|---|---|---|---|---|---|---|---|---|---|
| 1. | a | b | c | d | 21. | a | b | c | d |
| 2. | a | b | c | d | 22. | a | b | c | d |
| 3. | a | b | c | d | 23. | a | b | c | d |
| 4. | a | b | c | d | 24. | a | b | c | d |
| 5. | a | b | c | d | 25. | a | b | c | d |
| 6. | a | b | c | d | 26. | a | b | c | d |
| 7. | a | b | c | d | 27. | a | b | c | d |
| 8. | a | b | c | d | 28. | a | b | c | d |
| 9. | a | b | c | d | 29. | a | b | c | d |
| 10. | a | b | c | d | 30. | a | b | c | d |
| 11. | a | b | c | d | 31. | a | b | c | d |
| 12. | a | b | c | d | 32. | a | b | c | d |
| 13. | a | b | c | d | 33. | a | b | c | d |
| 14. | a | b | c | d | 34. | a | b | c | d |
| 15. | a | b | c | d | 35. | a | b | c | d |
| 16. | a | b | c | d | 36. | a | b | c | d |
| 17. | a | b | c | d | 37. | a | b | c | d |
| 18. | a | b | c | d | 38. | a | b | c | d |
| 19. | a | b | c | d | 39. | a | b | c | d |
| 20. | a | b | c | d | 40. | a | b | c | d |

# Answers

## These answers are given for guidance and are not necessarily the only possible solutions.

Where answers are not shown your answer will either be an individual one or shown in the text immediately preceding the question

## Chapter 1

p.9   See Figure 1.10.

p.10  SAQ 1. The solution would typically include: hours of work, holiday entitlement and pay arrangements, periods of notice, grievance procedure, health and safety at work, sickness entitlement, pension details
2. Any three of: the scope of work, the type of employee, the class of work, demarcation, method of working, wages, travelling and lodging allowances 3. Similar to that shown in Figure 1.10. 4. The solution should include: establish their identity and authority, maintain a record of who is on the site, advise them of the Health and Safety Rules which apply and any safety equipment required (instruction or information must be clear and easy to understand preferably in a written format), provide an escort to accompany the visitor where particular hazards exist. 5. The solution should include at least the following: where the work is to be carried out, when it will be started and when it is to be finished, exactly what is included, what standards it will conform to and which wiring system will be used.

## Chapter 2

p.11  contract of employment, skill, qualifications, experience, client, management contractor, sub-contractor, working relationships, requirements, contract, completion

p.19  Try this: The solution should include the following: Health and Safety at Work Etc. Act, 1974, The Management of Health and Safety at Work Regulations 1999, The Construction (Health, Safety and Welfare) Regulations 1996, The Construction (Head Protection) Regulations 1989, Lifting Operations & Lifting Regulations 1998, The Construction (Design and Management) Regulations 1994, The Provision and Use of Work Equipment Regulations (PUWER) 1998, The Personal Protective Equipment at Work Regulations 1992, The Manual Handling Operations Regulations 1992, Factories Act 1961 and the Regulations made under this act, Construction (General Provisions) Regulations 1961, Control of Substances Hazardous to Health Regulations 1999 (COSHH), Noise at Work Regulations 1989, Workplace (Health, Safety & Welfare) Regulations 1992, Electricity Supply

Regulations 1988, Electricity at Work Regulations 1989

p.25  SAQ 1. As the answer to the Try this p.19.

p.26  SAQ 2. The solution should include: design, construction, finance, resources, programme, technical requirements, legislation.
3. Increase in cost due to additional resource, time, material, delays to programme and failure to complete on time. Programme changes to accommodate the additional work, installations procedures, access and making good following changes.
4. The solution should include the following steps: identify the circuit to be isolated, check the voltage test instrument, check the circuit to be isolated is live, isolate the circuit, test the circuit is dead, check the voltage test instrument again, secure the isolation.
5. The solution should identify:
the areas of risk are controlled, activities are planned and monitored, minimises the danger to operatives and others, defines the area of work and activity precisely.
The solution may include:
working on or adjacent to live equipment, working in confined spaces, working on large and complex installations, work on part of an installation which is under the control of others.

## Chapter 3

p.27  Statutory, Non-statutory, employees, ignored, accidents, danger, Health and Safety Executive, permits, method.

p.35  Try this: The solution should include at least the following: 2nd fix materials: 1 × 1 gang 1 way switch, 8 × twin fluorescent fittings, 1 × Fire alarm bell, 2 × break glass call points, Fire alarm panel (case fitted at 1st fix), conduit box lids (option) Special tools and access equipment: (in addition to the standard electrician's tools, power drill etc.) MIMS cable: stripping tool, pot wrench, crimping tool; pairs of steps and access boards (option of small mobile scaffold); conduit bender and pipe vice

p.37  Try this: fan heater, switch with pilot light, dimmer switch, switched socket outlet with pilot lamp

p.38  Try this: battery, time switch, 2 wire series motor.

p.45  SAQ 1. (a) time sheets, daywork sheets, job sheets; (b) orders, delivery notes, invoices; (c) programmes and bar charts; (d) layout, circuit, wiring and block diagrams; (e) record drawings, as fitted drawings and overlays.
2. The solution should include: the work that is to be carried out, the address, customer's name, the date on which the work is to be done, any special instructions such as collecting keys, any special conditions that exist or equipment in use. It may include details of the materials to be used and the type of accessories.

3. The bar chart should contain the activities for the following: distribution board, fire alarm distribution board, trunking installation, conduit installation, MIMS installation. Consideration that more than one activity may progress at the same time should be recognised.

p.46 SAQ 4. The requisition should include: trunking, 90° trunking bend (internal), end caps for trunking, conduit, conduit couplers and bushes, conduit boxes (through and terminal), conduit saddles, switch box (surface metal clad), break glass points, MIMS cable, MIMS cable clips, MIMS terminations and shrouds, distribution board for power and lighting, distribution board and fire alarm, screws, wall-plugs and sundries.

5. The list should include layout drawings, circuit diagrams, wiring diagrams, record drawings and may include block diagrams, overlays and "as fitted" drawings.

# Chapter 4

p.47 control, monitoring, daywork sheets, activities, delivery notes, materials, circuit, wiring, block, record drawings, bar charts, Programming, essential, adequate, labour, progress.

p.52 Try this: (a) laser level or maybe a water level; (b) plumbline laser; (c) measuring wheel; (d) range finder or digital estimator

p.54 Try this: The solution should include: the company carrying out the survey, the owner of the property, the occupier of the property, any affected third parties.

p.55 SAQ 1. Authorisation to carry out the survey and appropriate insurance for the work and possible consequences.

2. The solution should include: measuring equipment (tapes, levels etc.) access equipment (steps, ladders etc.) means of recording the data collected (notepads, schedules, drawings etc.) and possibly the existing layout/record drawings

3. The solution should include: access to the site, the structure of the site, the material used for the construction, any third party requirements

4. The solution should include: determining cable length, size and type and calculation of voltage drop etc., access for plant, material and equipment, special requirements relative to the particular installation, extent of preparatory work, access equipment, tools and labour, quantities of material and equipment.

pp.56 and 57, Progress Check:
(1) c; (2) d; (3) b; (4) a; (5) c; (6) c; (7) b; (8) c; (9) a; (10) c; (11) a; (12) b; (13) c; (14) c; (15) d; (16) c; (17) a; (18) c; (19) a; (20) b

# Chapter 5

p.59 insurance, authorisation, extent, visual, measurements, structural, plant, measuring, tapes, wheels, levels, plumblines, range finders, access, working platform

p.62 Try this: 1. IP22 or IP21 2. Any application where protection against objects greater than 2.5 mm diameter/thick and where the equipment is likely to be subject to drops of water falling vertically.

p.65 Try this: 1. Any three from: flammable dust, vermin, corrosion, heat generated, mechanical protection requirements. 2. The solution should consider: physical interference from livestock and rodents, attack by chemicals, acid attack from animal waste, temperature and risk to livestock and staff from indirect contact.

p.72 SAQ 1. The solution should consider: steel flexible conduit, fibre flexible conduit, MIMS anti-vibration loop.

2. System selected should be SWA cable as this is flexible enough to be drawn through ducts, can be suspended at high level either fixed direct or attached to cable tray. MIMS may be considered as an alternative for similar reasons. Cost and availability may be the deciding factors.

3. The solution should consider: ambient temperature, changes in temperature, length of run, external influences such as exposure to sunlight, building movement, capacity of conduit and the number and size of conductors and mechanical protection.

4. The solution should include from the following:
(a) mild steel (b) copper, aluminium, gold, silver (c) PVC, glass, porcelain, some oils (d) copper, brass, steel (e) fibreglass, wood, polystyrene, rockwool.

5. The solution should include from the following: close proximity to equipment and machinery, sources of heat and cold, close proximity to electrical equipment, difficult access, gas or fumes, dust or fibres, fire or explosion, inadequate ventilation, poor quality of air or risk of air contamination.

# Chapter 6

p.73 equipment, materials, installation, corrosive, impact, solar, electromagnetic, strength, toughness, compatible, environment, roofs, confined.

p.84 SAQ 1. The solution should include: The Electrical Installation Certificate, used for certifying new electrical installation work including alterations and additions to existing electrical installations, The Minor Electrical Works Certificate, used for certifying alterations and additions to a single circuit – not to be used for the provision of a new circuit, The Periodic Inspection Report used to report on the condition of an existing electrical installation which has been energised and placed in service.

2. The solution should include: licensing, insurance purposes, sale of property, statutory responsibility, proposed development or extension to existing installation.

3. The solution should state: those responsible for the design, construction and inspection and testing of the installation.

4. The Electrical Installation Certificate should incorporate a schedule of items inspected and tested and a schedule of test results.

p.84    SAQ 5. The solution should refer to: the extent and limitations section of the report is only completed once the inspection and testing have been completed. The details agreed with the client may not be permitted by the users of the installation, or the purpose for which the installation is used. The true extent can only be detailed once the work has been completed.

## Chapter 7

p.85    inspected, tested, standards, certificate, Electrical Installation, alterations, additions, Minor Electrical Installations Certificate, new, existing, energised, placed, tests.

p.88    Try this: The solution should include: receive a request to tender, submit the tender, tender accepted, commence installation, complete installation and certify, request payment, payment received.

p.95    SAQ 1. The solution should identify the need for: a request to provide a tender is made, an offer is made in clear and concise terms, an unqualified acceptance of the offer is made.

p.96    SAQ 2. The solution should include: lapse due to time, withdrawal, death of the contractor, rejection.
        3. The solution should mention: the client, the architect, the consultant, the management contractor, the main contractor, the subcontractor.
        4. The solution should include: the earliest start date, delivery for items of special plant and equipment, availability of labour, availability of materials, access to site and work face, preliminary work, inspection and testing, commissioning.
        5. The solution should include: the bar chart which is used to plan and monitor progress of all activities on site. It is used to show the proposed work schedule and to allow actual progress to be monitored. The critical path which is used to identify and monitor those items which directly affect the completion of the project on time. Any change to the progress of the activities on the critical path will have a direct effect on the completion of the project. Activities which are delayed for too long may then become critical path items as they will have an effect on the project completion.

## Chapter 8

p.97    terms, conditions, tender, formed, accepted, consultant, monitoring, control, important, critical path, accurate.

p.106   SAQ 1. The solution should include the following: a request reference number, date the request was made, date by which a response must be received, subject of the request, date the response was actually received, whether the response was satisfactory, further action or request made (with the appropriate reference number).
        2. The solution should include the following: purchasing advantage, material control, the use of new technology where possible.

        3. The solution should include: cost of purchase, cost of hire, period for which it is required, further use, labour saving benefit, quantity.
        4. The solution should include: contract anomalies, shortage of resources, site constraints, environmental conditions, third party actions.
        5. The solution should include: planning, programming, monitoring, control of resources, best use of resources.

## Chapter 9

p.107   request, nature, reply, contracts, domestic, standard forms, information, contract, material, labour, purchasing, material, technology, purchase, planning, distribution, plant, anomalies. resources, environmental, programming, improve.

p.112   SAQ 1. The solution should include from the following: staff are clean and tidy, polite, helpful and knowledgeable and on time for appointments; staff treat customers with proper care and consideration, take the right tools, materials and equipment to the job; the company office is clean, tidy and welcoming, the workshop tidy and well kept, the company van and plant clean and in good condition; accurate records are kept and readily available, messages and information are relayed and enquiries dealt with efficiently, effectively, correctly and politely.
        2. The solution should include from: specification, operation and maintenance manuals, manufacturer's information, forms of certification.

p.113   SAQ 3. The solution should include: the number of points provided in each room, lights, sockets, telephone, TV and the like, the make and standard of accessories to be used, the extent of the work to be undertaken, any additional items such as outside lights, any exclusions like rewiring the central heating system or light fittings supplied by the customer, cost and programme.
        4. The solution should include colleagues and other contractors.
        5. The solution should include: inefficient working, disruption, additional cost, delays to programme, client dissatisfaction.

pp.114, 115 and 116 End Test:
        (1) b; (2) c; (3) b; (4) b; (5) d; (6) d; (7) c; (8) c; (9) a; (10) d; (11) b; (12) b; (13) c; (14) a; (15) d; (16) a; (17) b; (18) a; (19) a; (20) a; (21) b; (22) b; (23) b; (24) b; (25) d; (26) c; (27) d; (28) a; (29) b; (30) a; (31) c; (32) b; (33) b; (34) b; (35) c; (36) c; (37) b; (38) b; (39) a; (40) d